Outwitting Tomorrow

For a complete list of other books, audio and video tapes by Dr. Frank E. Stranges, write to:

IEC - Book & Tape Dept.
P.O. Box 73
Van Nuys, CA 91408-0073
United States of America

Visit Dr. Stranges on the web at www.nicufo.org

Dr. Frank E. Stranges, Ph.D.
Valiant Thor, Advisor

Outwitting Tomorrow

Universe Publishing Company

1st printing

Published by
Universe Publishing Company
P.O. Box 15655
North Hollywood, California 91615-5655

Library of Congress catalog card number: 00134729

ISBN 0-9655786-1-5

Advisor to author, Valiant Thor
Edited by Julie Corcoran
Cover concept & design by Jay Zeballos/J&L Advertising

Printed and bound in the United States of America

Table of Contents

Preface

Outwitting Tomorrow is intended for men and women in all walks of life who are striving to expand their lives and attain independence from those circumstances which are keeping them from experiencing the joys and happiness of living "the good life." By knowing and practicing a few simple rules and secrets and by doing what you can with what you have, wherever you are now in your own life, it is possible to bring about great results and new changes in your heart, soul and affairs, regardless of your age, education, health, environment or financial circumstances. These will thoroughly astound you.

The names of the figures who appear in *Outwitting Tomorrow* are of no particular significance. The characters themselves, however, represent the typical characteristics of each individual person who has started his/her walk on the Upward Path. One faces the same types of problems that perplexed John Miles, and we watch as he advances and emulates our Mr. Whitehead by applying his counsel to his own self just as we can do the same for ourselves. Then as he progresses still further, he takes on the character of still higher personalities. It is the sincere wish of the author and advisor that every person who reads *Outwitting Tomorrow* will begin at once a transition from the John Miles stage, which represents the masses, to that of Mr. Whitehead, who is representative of the liberated individual and then beyond to what Destiny has intended.

Outwitting Tomorrow gives you a simple, easy-to-follow plan whereby you may master your future instead of allowing the future to dictate to you. For if you are controlled, you are merely a puppet of fate; but if you are the master, you will travel the bright Upward Path that Destiny has prepared for you.

The contents of *Outwitting Tomorrow* have been garnered from the four corners of the globe. The author regrets that, due to the length of the list and lack of space, he cannot give due credit to each one of those individuals who contributed in some small way to this work. He must satisfy himself with the assurance of his deep appreciation to each person who had even a small part in the production of this book and with the dedication of the book to the Individual, whoever and wherever he or she may be.

CHAPTER 1

The Future Unfolds

It was late afternoon when John Miles, carrying a pack and breathing heavily, reached the sharp crest of a low ridge of mountains facing the Pacific Ocean. As he emerged wearily from the shadowy black slope, shafts from the low-sinking sun dazzled him with their splendor. He drank in the colorful sight for a moment and then, with his eyes still fastened upon it, lowered himself to a convenient rock with a sigh of relief and satisfaction.

"Beautiful sight, isn't it?" observed a voice close by.

With a start of surprise, Miles turned toward the voice and saw a man seated on a ledge a few feet away. Evidently in his haste to reach this resting place he had not noticed that another person was already there, viewing the sun ready to dip behind the vast Pacific, flashing to shore a broad path of shimmering gold across the dark blue water.

"Oh, it's pretty enough," Miles said, in the manner of one who has witnessed the same spectacle hundreds and hundreds of times, "but when you see it like that, there's some nasty weather on the way."

And so, from this banal subject they gravitated into an impromptu conversation. It developed that Miles had moved to this place from an Eastern industrial city, having inherited the property from a distant relative. Although a full section of land, it was of no great value. Most of it he had leased out to sheep-raisers for grazing land. But just over the ridge was a small, fertile valley, and this was the property that he began to cultivate many years ago. Someone, probably an early Californian, had built a Spanish adobe house there. It had a red tile roof. The thick, sun-baked brick walls kept it warm and snug in the winter and cool in the hot summer. One corner of the section of land extended over the mountain and down to the sea, and here, where it met the highway, was erected a fairly large, well-built residence where the Miles family lived.

The stranger's name was Whitehead. He had just come out here from the midwest. He said he was on a business trip and was stopping at the hotel in Casa Del Rey. He was well-dressed, tall and his face showed fine, strong features. It was plain to see that he was a man of purpose, capable of engineering a well-organized plan of action.

"Quite a bundle you have there," he noted, eyeing Miles' overflowing backpack.

"It is that," Miles assured him. "You see, I grow fruit and vegetables on my ranch down there." He pointed to the valley from which he had just climbed. "And almost every evening I take something home to my family. Down there," waving a hand toward a house on the highway near the ocean, "is where I live."

"Surely, you don't have to haul all that you grow all the way over this ridge," protested Mr. Whitehead.

"Oh, no," answered Miles. "There is a road which comes in at the other end of the valley that I use for transporting the fruits of my labor. However, it is 12 miles by the road and only three-quarters of a mile over the ridge, so I usually hike over."

"I see," said Mr. Whitehead. "You save time and in addition you get a lot of good exercise."

"I could do without the exercise," replied Miles. "When I was younger I didn't mind so much. But now rheumatism and stiff joints have made it a painful climb."

"I didn't find the climb to the summit particularly hard," said Mr. Whitehead.

"Well, you wouldn't," Miles said, rather shortly. "You're a tourist, you see, and you climb up here for the fun of it. Besides, the ocean side of this ridge isn't nearly as steep as the other side. Anyway, you look like you're a much younger man than I am."

"You're partly right, my friend," Mr. Whitehead agreed. "I climb mountains for the fun of it. In fact, everything I do is for the fun of it. I haven't had a stiff or painful joint in my body for years. But that last remark about my being older than you; well, I seriously doubt that. How old are you, if I may ask?"

"Sixty-five this coming November," came the prompt reply.

"Well," Mr. Whitehead pondered, "you hardly look that old. You still have a lot of good years ahead of you."

"Oh, I'm in fairly good condition," Miles returned, "but I've worked hard all my life and there's nothing like hard work to cripple a man up and make him old. Believe me, though, when I was your age I was as strong as an ox and never knew what it was to be sick or under the weather."

"You say when you were my age. How old do you think I am, anyway?"

"About forty-five, I should say," Miles estimated, squinting his eyes in what was left of the sunlight. He was sorry the moment he spoke because it instantly struck him that Whitehead looked hardly even forty.

"Well, you're way off my friend. So far off that you'd need additional guesses to come anywhere near the truth. Actually, I'm past seventy. How far past is another guess for you."

Miles viewed the speaker with skeptical amazement. Then, as he realized that the man was speaking earnestly, he attempted to justify his estimate.

"Yes, you may be. But you see, I've had to work hard all my life!"

"So have I," replied Mr. Whitehead quickly. Then softening his voice he continued, "at least I did until I was about your present age; then I started to play. I don't mean that I retired. I mean that up until then everything I did was work. Then, one day, I changed my thinking and everything I did from that time to now became play. I found that when we hate our work, even if it's the lightest kind of a job, it becomes drudgery, and we grow old under its burden. But when we love our work, then it is no longer work, but play. The harder we play, the younger we grow."

Miles held his tongue, but his thoughts about Mr. Whitehead were not complimentary. He said to himself, "He's deliberately lied about his age and now he says it's possible to work and call it play and like it. Either he's a down-right liar or else he's filled with harebrained ideas."

Whitehead broke the silence with, "I'm neither exaggerating nor am I mentally unbalanced. What I have told you is simply the truth."

Miles' face turned deep red with embarrassment. He knew that the stranger had read his mind and he fumbled about for an appropriate apology.

"Oh, that's all right," Mr. Whitehead cut in. "You've got the right to think anything you like, but with me you can think it out loud whenever you want. I like men to be straightforward with me."

His voice rang with such perfect sincerity that all doubt of what he had said regarding his age and his ability to make work play was instantly swept from Miles' mind and he regarded Whitehead with respect.

By this time the setting sun was barely visible over the horizon and its fading rays cast a rich, mellow glow over the edge of the sea.

"In less than a minute," Mr. Whitehead said quietly, "the sun will be out of sight. Take special notice a few seconds after it disappears and tell me if you can see the place on the horizon where it has just been light up for a moment."

Both men watched intently. The sun, with a final dip, disappeared from view and in a moment Miles exclaimed, "Look there! Look at that brilliant flash! Now that's something remarkable. Imagine me watching it set for all these years and never noticing that before."

"Well, you are to be congratulated," Mr. Whitehead replied. "You know, many people can never see it."

"But how do you account for it?" Miles asked.

"There are a lot of things I don't even attempt to account for now," Mr. Whitehead answered. "That's one of them. Some day, however, I will find out."

The darkness now was closing in rapidly and Miles ventured, "We'd better get on our way or it will be completely dark before we get to the highway."

Mr. Whitehead nodded in agreement and rose from the ledge

where he had been seated. They started down the trail together, both men proceeding rapidly. Miles had evidently forgotten all about his rheumatic joints, so interested had he become in the other man.

Reaching the highway, Miles started down the road homeward, but Mr. Whitehead detained him saying, "I have my car here. Let me drive you home." And almost before he knew it, Miles was riding in the most expensive car he had ever seen. A few moments later, he was getting out in front of his home and thanking Mr. Whitehead for the lift.

"Will I see you again before you leave our section of the coast?" he inquired anxiously.

"I'll be here for a few weeks yet, and I'll watch another sunset with you soon," Whitehead replied graciously. "Good night."

In a moment the powerful car was sailing down the highway and soon was out of sight as it turned along the winding coast road.

CHAPTER 2

Observe and Believe

Miles was so occupied with his thoughts at the dinner table that evening that he entirely neglected to make his usual grumbling complaints about the food, although Mrs. Miles was an excellent cook. This was most amazing to her and the children who were nearly grown. They could not remember his ever having eaten an entire meal without grumbling some kind of disapproval over something.

He retired early that night and fell asleep still meditating on his conversation with the stranger.

On his way to the ranch the next morning, he sat down to rest on the same rock he had occupied the evening before and tried to reconstruct the scene but there was a vast change. The gorgeous ocean of yesterday was now a sickly looking green and the golden sky had become shrouded with a cold, gray mist. The sun was a palely reflected light toward the southeast. His thoughts again turned to Whitehead and his youthful appearance for a man over seventy, his air of mystery. Yet he was warm and sociable, and was extremely fascinating.

After a few minutes Miles started the descent to the ranch and although it looked drab and uninviting in the wan, morning light, he did not mind it so much. A great indefinable change was taking place within him.

The Miles family had somehow fallen into the habit of referring to the adobe structure in the valley as the "hut." It consisted of a spacious living room, a bedroom and a kitchen which also served as a dining room. It was still in a fair state of preservation despite its age. The "hut" had become a sort of sanctuary for Miles where he could read his books and papers, and in which he had spent many long, delightful hours alone contemplating and meditating.

Each summer, however, the old place came to life again. When the berries and the orchard fruits ripened, Mrs. Miles descended upon it with the children, and there would ensue a busy two or three weeks of cooking and canning and preserving.

Upon arriving this particular morning, Miles selected garden implements from the tool shed and was soon industriously engaged digging in a patch of tomato plants. He worked with a strange new zest, determinedly ridding the plants of the obnoxious weeds which threatened their growth. As he worked he seemed to hear again the voice of Mr. Whitehead saying, "If you like work, then work isn't work, it's play." And within him arose a gleeful feeling of being light on his feet which made him exceedingly happy.

After a long while thus engaged, he paused and straightened up. Looking at his watch he discovered, much to his amazement, that he had worked clear through the lunch hour and so he made his way to the "hut", started his coffee, and ate with great delight the lunch Mrs. Miles had left for him.

The afternoon passed quickly and as the long shadows started climbing the opposite slope, almost regretfully he gathered up his tools and prepared to leave for home. As usual, he stopped at his favorite rock at the summit to rest, and as the sun sank below the golden line of the horizon, he waited expectantly for the "flare up." Sure enough, there appeared a momentary recurrence of the sun's brilliance and with a thrill he recalled that Mr. Whitehead had said only a few people could see it. It gave him a feeling of superiority and an exaltation and refinement of spirit to realize that he was capable of seeing something which few men could know. With great joy he descended the slope never realizing that, for once, the pangs of stiff joints and rheumatism had excluded him.

A week went by and nothing new occurred, yet Miles' strange new feeling persisted and each morning he eagerly looked forward to the day's work. Another week passed and then one evening, just when he was beginning to think that Mr. Whitehead had deserted him, the same long, expensive car appeared before the gate. There at the wheel with a smile and a cheery greeting sat Mr. Whitehead.

Declining Miles' earnest invitation to come in the house, he explained that he was just returning from a trip to the north and he must get back to his hotel at once. However, if it was agreeable to Miles, he would like to spend the following day with him at the ranch. After a fervent and heartfelt assurance on Miles' part that nothing would please him better, they agreed on a time to start for the ranch in the morning and Mr. Whitehead went on his way.

The following morning the two men set off for the ranch, fortified with a splendid lunch which Mrs. Miles had prepared for them.

Upon arriving, they started to work at once. Miles learned that Whitehead was no idle rich man, but one who knew agriculture thoroughly and could do more than his share of the work quickly and efficiently. At one o'clock they put away their tools, having accomplished in a half day more than Miles, alone, usually did in two days. They proceeded to the "hut" for lunch. Coffee was soon ready and the men began devouring Mrs. Miles' lunch with an appetite enhanced by good, honest toil and they consumed it to the last crumb. Completely satisfied, they relaxed in easy chairs in the living room. Finally, the silence was broken by Mr. Whitehead.

"Looks like the children used the wall over there as a blackboard," he observed, pointing to a number of drawings across the room.

"Oh, the kids used to make themselves at home around here," Miles replied. "Now, since they think they're grown up, it's about all we can do to get them to come here for a couple of weeks in the summer to help their mother can fruit."

"Whoever did those drawings was no average person," commented Mr. Whitehead.

"My son, Michael, drew them some time ago," Miles told him. "He's different from the rest of the family, kind of an odd young man. He's working his way through college right now and in a few weeks, when school is out, I imagine he'll come home for a visit."

"Where did he get all his ideas for these drawings?" Whitehead questioned, interested.

"Oh, I don't know. Maybe he saw them in a book somewhere. He's the bookworm of the family," replied Miles.

Whitehead, who was busy studying the drawings, did not seem to hear Miles' last remark. "I must say, that for a mere youngster, that large middle sketch is quite a masterpiece," he thought to himself. "Let's see, the past, present and future are all represented in these lines." Then aloud he said, "Miles, I wonder if you know just what that large triangle represents?"

"Michael called it 'the Pyramid'," replied Miles. "He used to stare at it so much that sometimes I had to talk pretty sharply to him. He'd actually go without eating just to study it."

"I can't say that I blame him much," commented Mr. Whitehead. "This is a sketch of the Great Pyramid of Gizeh and the lines which you see within its borders indicate the rooms and passageways it contains."

Mr. Whitehead arose and walked over to the drawing to view it more closely. "Every part is drawn to exact scale!" he exclaimed, softly. "That boy of yours is certainly all right. I hope I have the pleasure of meeting him in person some time."

Miles was speechless. "Why," he stammered, "I thought it was a

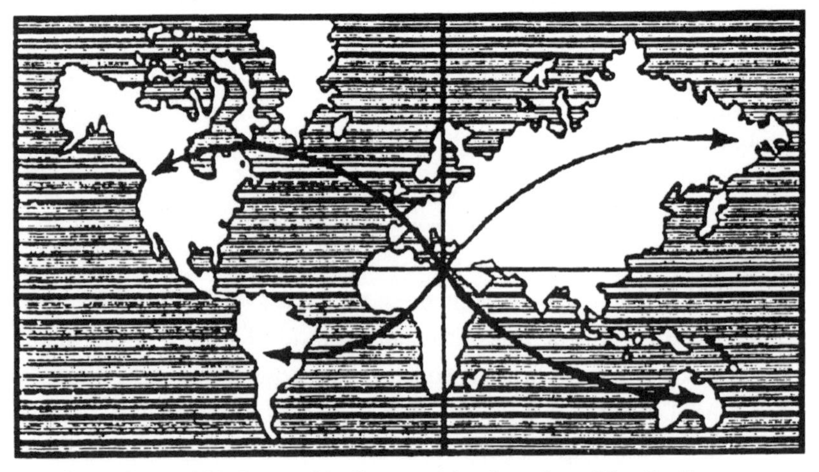

The Great Pyramid is located in the exact land center of the Earth.

foolish drawing. I didn't know it meant anything at all."

"It means much more than you think," replied Mr. Whitehead. "That drawing is a chart of time and events. It represents approximately 6,000 years in the life of mankind. It dates back to 4,000 B.C., and then it proceeds to the beginning of the Christian era, then forward in time, and on to the end of this millennium. In this period of 6,000 years the most important events of human history take place. The most amazing changes, however, are made between the present time and the beginning of the grand, New Millennium—the year 2,000 A.D."

Miles just sat and stared. It was incredible that such a thing could have been right before him all this time without his knowing its significance. Why hadn't Michael told him?

Finally he said, half to himself, "Well, what do you know about that? I'm glad there's someone in the Miles family who is interested in something more than just the ordinary things of day to day life."

CHAPTER 3

Designs of a Lifetime

Miles arose and hurried over to the drawings with a newfound interest. "If Mr. Whitehead considers it so important there must be something to it," he thought. "And Michael, he must be more than just an ordinary young person to have drawn this. What could it all mean? And what was this talk of the future as well as the past and present? Did he mean that the drawing actually told about the future?" By now his curiosity was truly aroused.

"It's all very strange to me, Mr. Whitehead," he said uncertainly. "Would you explain to me what this Pyramid drawing is all about?"

"It would require a number of very large books to tell all its significance and its wonders," replied Mr. Whitehead. "Just relating the highlights to you would be quite a long story and, after all, the garden needs some attention. Perhaps we had better wait until another day to get into it."

"There isn't a thing in the garden that needs immediate attention. Tell me about the Pyramid now," Miles almost demanded.

His eagerness brought a smile to Whitehead's face and to his eyes, and settling back in his chair, he said, "So be it. But before we get into the purpose of this grand structure, I think it would be better for you to first know something concerning its age, size and location.

"To start, it was built close to the River Nile in Egypt and is referred to in Sacred Scripture as an 'altar' and also, as a 'pillar': 'In that day shall there be an *altar* to the Lord in the midst of the land of Egypt, and a *pillar* at the border thereof to the Lord. And it shall be for a Sign and for a Witness unto the Lord of Hosts in the Land of Egypt.'

"True enough, the Great Pyramid is located in the land of Egypt, and further than that, on the border of Egypt. It was built on the west side of the Nile, high above the river. The base of the Pyramid covers more than 13 acres.

"It was one of the largest jobs ever undertaken by man," continued Mr. Whitehead. "I say undertaken by man, but actually, there is no indication that man as we know him today ever built it. In fact, there is no authentic source in this material world to which we can go for accurate information regarding the origin or the builders of the Great Pyramid. It has been assumed for all these years that slaves built this grand structure, but there is no recorded history available to mortal men to prove that this is true. Many archaeologists have proposed their own theories of how it was done, but none seems to quite explain the entire truth of it all.

"Thousands of the stones used in its construction are perfectly square and as tall as an average man with some of them weighing over 30 tons. When these huge stones were all in place, the outside of the Pyramid resembled a staircase.

"Finally, the outside was made perfectly smooth by placing on these steps beveled blocks of limestone which were so perfectly fitted together, and the whole structure was so finely polished, that it would

"In that day there shall be an altar to the Lord in the midst of the Land of Egypt, and a pillar at the border thereof to the Lord. And it shall be for a Sign and for a witness unto the Lord of Hosts in the Land of Egypt."
— Isaiah 19:19,20

have been impossible to introduce the point of a very thin knife blade between them. In fact, you could not see at a short distance where these gigantic blocks were joined. When you consider that each of the four sides of the Great Pyramid was 5¹/₂ acres[1] in area, you can easily believe that this was the largest polishing job ever undertaken and completed.

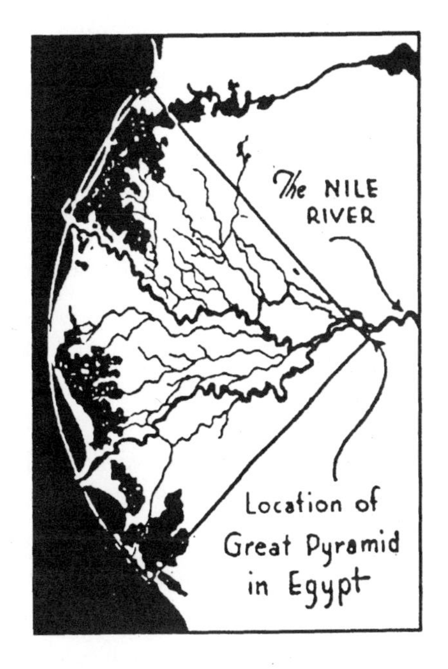

"When viewing the Great Pyramid, I often picture to myself young Egyptian boys climbing up the sides a ways and then 'sliding' down the highly polished surface in one fell swoop landing on a freshly blown sand bank. Just how they would climb it would be a problem, but leave that to the boys."

Mr. Miles sat spellbound. As Mr. Whitehead paused, Miles said, "I have seen pictures of the Great Pyramid, but I had never really stopped to consider its massive size."

"It is, as you know, one of the Seven Wonders of the World," replied Mr. Whitehead. "It could not be duplicated today. There is no known method by which those huge stones could be cut so perfectly and no modern machinery is capable of fitting them into place so precisely."

"There is nothing to indicate when the Great Pyramid was built," continued Mr. Whitehead, "although it is generally thought to have been around 4,000 years ago. Some scientists, however, claim it was built nearly 6,000 years ago. Who built it is equally unknown. There are many theories regarding the builders and the methods used, but nothing authentic, nothing more than guesses."

"In that case," said Miles, "No one knows just why the Great Pyramid was built either."

"Until recently," answered Mr. Whitehead, "no one had the slightest idea why it was built. It was thought to be just a huge tomb for an Egyptian Pharaoh. It was found, however, that no one was buried there; and so another reason was sought.

"At last, after many exact measurements, it was discovered that the dimensions of the ancient passageways and rooms had a common connection with known time, measurements and quantities.[2] For example, in one instance the exact length of the year was given; then the exact diameter and weight of the Earth. More and more discoveries were made until now we know beyond a doubt that those ancient builders knew far more than we know today about everything under the sun. Even squaring the circle was done with such ease and simplicity it was considered nothing to them."

"This is truly remarkable," said Miles, "but these wise ancients must have had some reason for building such a huge structure. Is that reason known to us today?"

"As far as we know, there is only one for its construction." answered Mr. Whitehead. "It is believed that it was built for the express purpose of enlightening men and women the world over. Although it deals with mankind from 4,000 B.C. up to the present time, its real purpose was intended only for certain members of the human race to use for purposes of increasing understanding of the spiritual aspects of the universe from, say, the beginning of World War I until some measure of time which will take place after the year 2,000 or 2,001 A.D.

"It should be noted now that the The Creator listens to the prayers of his people, especially those who have begun to 'turn West' and have started on the Path of Expansion. We all know that prayer changes things. Time has been changed during recent years because of the prayers of the faithful who have continued to stay the course which Destiny has set for them and they continue to expand and teach others. This is why particular dates in the future are subject to change. Devotion, worship and obedience to the universal laws of creation are part and parcel of this reasoning.

"In order to explain to you the prophecies of the past, how they have been fulfilled and are being fulfilled and just what is to come, I must give you a glimpse of the interior of the Great Pyramid. It is the interior that is important to those interested in progressing on the Path of Expansion."

CHAPTER 4

The Progress of Time

"Your son, Michael," continued Mr. Whitehead, "did a very thorough job in designating the rooms of the Great Pyramid. He also numbered the passageways so simply that we'll have little difficulty if any making our way through the huge structure. We must first ascend the North side of the Pyramid about fifty feet to the entrance. Once inside, we find ourselves in Passageway No. 3, which is only $47^{1}/_{2}$ inches high, so stoop down or you'll bump your head.

"Traveling along the downward slanting floor for about 90 feet, we find the place marked XX, with another passageway branching off on an upward slant marked No.2. Although the one we're on keeps descending, we will first explore this new one. It proves to be the same size as the first until, after climbing 128 feet, the roof suddenly rises from $47^{1}/_{2}$ inches tall to 28 feet. This section is known as the Grand Gallery. It is now possible to straighten up and walk.

"Having crossed this Grand Gallery for about 156 feet, we come to the Great Step which is exactly 3 feet high. Clambering up this, we find ourselves at the end of the Grand Gallery and confronted by another low passageway which is only 43 inches high. But, thankfully, we discover that it extends only 4 feet, 4 inches, where it terminates in a room which is about 13 feet high and over 8 feet long, called the Ante Room.

"We leave this through still another low passageway, 8-1/2 feet long, which takes us to the King's Chamber, the largest room in the Pyramid. It is practically 19 feet high, 17 feet wide and slightly over 34 feet long. Its walls are made of a beautiful rose red granite. Nine immense stone beams, which weigh on an average of 30 tons each, comprise its ceiling. Where these stones fit together, the seam is so minute that it can barely be seen at close range, probably the most perfect stone-fitting job ever known. The walls, floors and ceilings are built from exactly 100 stones. The King's Chamber is empty except for a lidless open treasure chest at one end. This chest must not be mistaken for a casket. It is not just a huge burial monument for some Egyptian king. In fact, it was not even built by the Egyptians, as were the other 37 pyramids in Egypt.

"Let's now go to the Queen's Chamber. Retracing our steps through the low passageway, the Ante Room and the beautiful Grand Gallery, we come to the place marked X. At this point another long, low passageway, No.4, extends back in the direction from which we just came, but this one is absolutely level. It's about 130 feet to the Queen's Chamber which is situated directly under the peak of the Great Pyramid. Although not so large as the King's Chamber, it's very beautiful. Its gabled ceiling

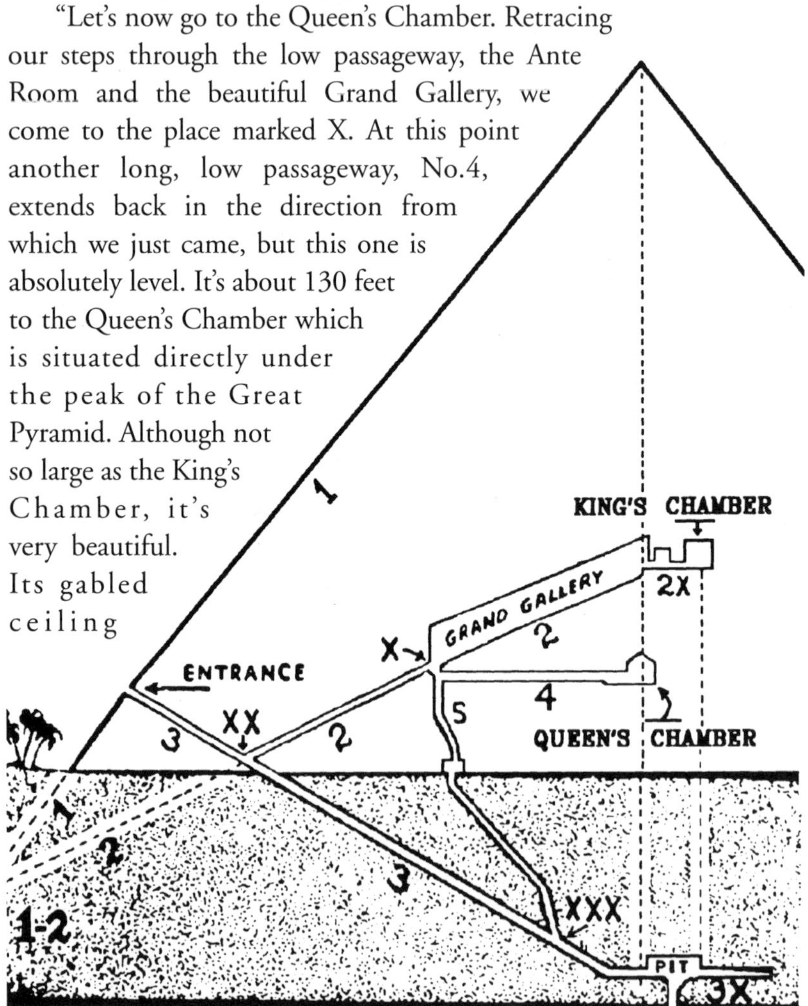

rises to a trifle over 15 feet in height.

"Again retracing our way, we return to the point marked XX. Here, if you remember, we left Passageway No.3 to ascend Passageway No.2 and explore the King's Chamber. So now let us continue down Passageway No.3 until, far below the base of the Great Pyramid, 325 feet from where we entered it, we arrive at the Bottomless Pit. You will hear more about the 'Pit' later. For now, I would warn you that it would be dangerous here without a light. Right in the middle of this room is a jagged hole large enough into which a man can plummet and which runs straight down to a considerable depth.

"On the far side of the room is another low passageway, marked 3X, which goes forward 54 feet and comes to a dead end. Its size dwindles as it progresses so that its far end is much smaller than the end which begins in the room known as the Bottomless Pit.

"From here we go back to the place marked XXX. Here we find a steeply inclined passageway, No.5, which leads to a small, unfinished room known as the Grotto, which serves as a resting place before we continue our upward climb. From here on the passageway is perpendicular, but at last we arrive once more at X, and from here we go back down to XX, and then up Passageway No.3 to the entrance and out into the sunlight.

"And so we have been through all the rooms and passageways of the Great Pyramid covering a distance of approximately 1,300 feet. What a marvelous piece of construction! But the most fascinating part comes when we examine the reason for its construction.

"It seems reasonable to believe," he went on, "that such a huge structure must have been built for a very particular and important purpose. Although the Great Pyramid has been open since 820 A.D., it was not until shortly before World War I that scientists, who had thoroughly measured it in all its details, made the startling discovery that the dimensions of the rooms and passageways corresponded to points of time in world history. It then became evident to the scientists that the purpose of the Great Pyramid was to foretell or prophesy the times and conditions to come. According to the scientists, this prophesy begins in 4,000 B.C. and extends forward to approximately 2,000 A.D., a period of 6,000 years. But keep in mind that these were men interpreting what they saw as prophesy and not necessarily Divine Prophetic Information.

"In order to make this period of time plain to you, we will again refer to Michael's Pyramid drawing. By extending Line No.1 and Line No.2 downward, they eventually converge. This point of convergence, marked 1-2, represents the year 4,000 B.C. Then, disregarding Line No.1, each inch represents one year of time as we move up Line No.2. It sounds odd to figure this in inches but Pyramid inches are identical to American inches where only a trifling difference is discernible in 1,000 inches.

"According to this measurement, then, it develops that the point marked XX represents the time when the Children of Israel crossed the River Nile and came out of Egypt. Following the course of humanity up Line No.2[3] at the rate of one year per inch, we find that the Birth Of Christ, occurring between 1 B.C. and 1 A.D., is indicated by the entrance to the Grand Gallery. Here we have completed a 4,000 year period from our starting point.

"Finally, having almost reached the upper end of the Grand Gallery, we find ourselves at the Great Step. Having traveled 1,844 inches in the Grand Gallery, we find that this point represents the year 1,844 A.D. The Great Step is three feet high; but it doesn't represent 36 years in time, although it does represent a time period. It represents a time when mankind took a great step upward and time sped up.

"The entrance of Great Britain into the World War in 1914 (what was supposed to be the 'war to end all wars') was thought to be represented by the beginning of the low passageway just beyond the top of the Great Step, and that from the top of the step onward, time, having quickened, was represented at the rate of one month per inch. If this were true, then the end of the low passageway would indicate the the end of World War I. According to the new measurement, this would be on November 11, 1918. But during that year, another type of horror befell mankind.

"It is worthwhile to note that in early 1918, an influenza had begun in the United States. In addition, at this same time, one and one-half million U.S. soldiers set out on transports to enter the war. This was the largest relocation of people in recorded history. And many of them had already been infected with the flu virus.

"Meanwhile, at the army base at Ft. Riley, Kansas, soldiers began to burn large piles of manure and as a windstorm picked up fury, it filled the air with infected virus and soldiers by the thousands were exposed to the deadly virus. Many of them again packed into transports which were headed for France and the French soldiers began to fall ill, then British, then German (160,000 Berliners died in the epidemic), continued its spread across Asia, China and finally reached Spain where it then became known as 'the Spanish Flu'. The war proved a terrible propagator of this new virus. Men were falling dead on the battle fields, not from wounds, but from this unknown virus.

"On August 12, 1918, a Norwegian Liner entered the port in the United States with 100 sick passengers who scattered into the population. The virus had come full circle. Too small for anyone to see with the naked eye and mutating along the way, it began causing secondary bacterial infections which proved so very deadly. No family in the entire world was untouched by death from the flu. Chaos broke out as the systems of every city became overwhelmed by death. Scripture says that "He has set a boundary for the darkness", yet death was everywhere.

"By Autumn, 1918, the epidemic was raging, following the pioneer routes from the East coast to the West. No one was left untouched. Our excitement waned in view of these developments.

"All during the forenoon of November 11, 1918, we who knew of the prophetic significance of the Great Pyramid, anxiously awaited developments. Had we calculated correctly? Or had the prophesies ceased at the foot of the Great Step? As you know, we soon found out. While we were overjoyed that the horrible and ghastly debacle had ended, we were also greatly elated to know that the time measurements of the Great Pyramid had again been proved correct.

"The jubilation at the end of the war only served to increase the spread of the unstoppable Spanish Flu. 44,000 U.S. soldiers overseas had died and 5,000 sailors had been killed by this terrible menace. If there was a transport ship carrying 10,000, then 4,000 were lost to the flu. One-quarter million German soldiers died. Everyone on Earth had been exposed. Imagine it, every single person on the face of the entire planet was exposed! Yet not everyone died from it. Some people did recover to the amazement of the medical/scientific community and yet there was no way to understand why some survived it, what had caused it, how to treat it, how to prevent it. The virus is now being kept by the Center for Disease Control, as well as several other specially built sealed rooms in order to prevent escape of this deadly menace.

"At about the same time the war was ending, the virus just sort of began on its own to kill itself and in time stopped as quickly as it started. In the end, between twenty and thirty million people had died from this flu. This illustrates certain possibilities which can and have taken place in the history of mankind. Literally millions of people died during this horrendous period of time. Many others took place which destroyed millions including the Black Plague in Europe. And it could happen again, no doubt about it. This is why concern is high regarding the possibilities of germ and bacterial warfare.

"Now back to the drawing. Following the diagram again, we find that at the end of World War I we emerge from the low passageway into the Ante Chamber. Here the ceiling is nearly 13 feet high, plenty of room for one to stand up straight after being cramped in the low passageway. This Ante Chamber represents that period directly after World War I when everyone was prospering. But this is not to last very long.

"At the South end of the Ante Chamber you will see another low passageway, the lowest of them all. This indicated another trying time for the world, and many people were fearful that war would break out again. August 29, 1928, was the date signified for this new World War to start, and we were relieved when that day came and went bringing no war. But something did start right then and there. Something which has changed the attitude and destiny of Great Britain and the United States more than did World War I. On May 29, 1928, the greatest depression of all time started in England. A little later, as was the case in World War I, the same depression hit the United States when the stock market crashed.

"People who had never wanted for anything found themselves destitute. Solid, substantial, hard-working citizens were faced with accepting charity. Dark-skinned natives in Africa and South America who had never in their lives heard the names of Downing or Wall Streets slipped back into the endless jungles and reverted to a more primitive existence.

"The end of this last low passageway corresponds in time to September 16, 1936. On this date, England started emerging from the World Depression. Munitions factories had done the trick. War scares had finally forced her to start preparing for the eventuality of an attack, and so labor was in demand with an increased circulation of currency. The United States, late to enter the depression, was slow to recover from it. Being in a more secure position geographically, the necessity of a huge national defense operation was not so apparent, so that way out of their difficulties was not open to them at that time.

"Getting back to the diagram once more, you will see that we are now in the King's Chamber. You can imagine what a relief it would be if we were actually journeying through the Pyramid to get out of that low, cramped passageway and into this large, magnificent room where we could stand up, stretch and even jump up and down without bumping our heads. We encounter many significant dates as we traverse the King's Chamber. They are calculated by carrying imaginary lines from other points of the Pyramid so that they cut across the floor of the King's Chamber.

"Scripture tells us of a 'one-hour period'. As we journey across the King's Chamber, we come to a line which represents the beginning of this 'one-hour'. In reality, it is a 15-year period beginning August 20,1938 and ending August 20, 1953. Between these dates man experienced great changes.

"We are fortunate to be living now in a most interesting and valuable period. When I was younger and knew nothing of things to come, I often mourned the passing of the pioneer days. Pioneering and trail-blazing held a strong appeal for me. But after having learned my true mission in life and embarking upon this most intensely interesting and exciting work, it dawned upon me that after all, I was to satisfy my ambition to be a pioneer in a far different sphere of activity than I had imagined. Actual pioneering fascinates me more and more each day as its true meaning continues to unfold."

With an unspoken understanding, they both arose and prepared to leave.

For a moment Miles thought and then, carefully choosing his words said, "You have given me so much information, so much that I never heard before, that I hesitate to make any comment on it now. I need time to think. I will probably need to ask a great many questions. When I have digested what you have told me, I can discuss it more intelligently."

"Splendid," approved Mr. Whitehead. "That is exactly what you should do. 'Sleep on it' and tomorrow, we can discuss your questions. In the morning I must attend to some business in town. You go to the ranch early and keep busy, both physically and mentally, and I'll join you as quickly as possible."

Leaving Miles at his home, Mr. Whitehead continued on to Casa Del Rey.

CHAPTER 5

The One-Hour Period

It was eleven o'clock the next morning when Mr. Whitehead arrived at the ranch. Miles, happily busy in a bean patch, greeted him cheerfully, inviting him to pitch in and have some fun. Willingly, Mr. Whitehead removed his coat and soon was diligently weeding. Upon arriving at the end of a row where the vines were growing luxuriously, Mr. Whitehead inquired,

"Have you ever paid any attention to the manner in which these vines wind themselves around the poles?"

"Why, they just twist around as they grow," replied Miles.

"That's true, but the point I want to bring out is that they twist themselves around the pole in only one direction: counter-clockwise," Miles stated.

With a puzzled look on his face, Miles examined the bean plant nearest to where he was working. It wound counter-clockwise. He examined another, and another. They all wound counter-clockwise!

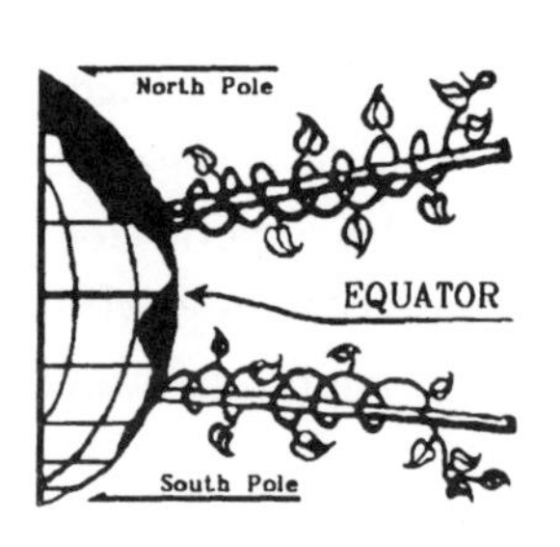

"I've been growing beans for years," he confessed, "but I never knew they grew that way before. Did you ever see one that wound the other way?"

"Some years ago, on a trip to Argentina, I first learned of this phenomenon. Two fellow passengers, one an American, the other an Argentinian, were discussing the relative merits of the beans in their respective countries. The American, eager to show his superiority, sought to confound the Argentinian by asserting that all bean vines wound themselves around their poles in a counter-clockwise direction. The Argentinian, equally positive, denied the truth of this statement, claiming that just the opposite was true. All bean vines wound themselves around clockwise.

"The argument grew in intensity and finally resulted in their making quite a heavy wager with the understanding that immediately upon arriving in Buenos Aires each would seek to prove his point. I was invited to go along to judge whose claim was correct. By this time I was so intrigued by their propositions that I agreed to do so.

"Upon arriving in port the next morning, we hired a taxicab and set out to find a bean patch. After considerable haggling with the driver we managed to impress upon him the importance of our little adventure. He drove us to a bean patch some distance outside the city. The American took a stand to one side and stubbornly challenged us to prove him wrong. We promptly made our way into the bean field. After a few moments we unanimously reported that every vine wound itself around the pole in a clockwise direction. I wish you could have seen the amazement and bewilderment on that American's face when he realized he was wrong. He just couldn't understand it. He had been so sure that he was right.

"Later on, after I had returned to the United States, I remembered the incident, and on a sudden impulse I decided to find a bean patch and see the curious sight of all the vines winding themselves clockwise.

"So this time I was alone and I set out for the countryside to find a bean patch. Behold! Every vine wound itself around the pole in a counter-clockwise direction. I felt very much as my American friend must have felt in Argentina. Upon further investigation I began to understand that both men had been right. North of the equator vines wind counter-clockwise, while south of the equator they wind clockwise. I also discovered that the same principle applies to water going down a drain. It was really quite fascinating.

"The only significance of the vines twisting one way or another is the application of the idea to humanity. The clockwise movement represents mass-minded humanity. They are perfectly satisfied to 'string along' with man-made ideas, opinions and traditions, all of which are decidedly negative. The counter-clockwise movement is just the opposite. It is positive and represents a new way of thinking. It constantly turns against everything negative and materialistic and represents the individual expanding into life.

"And now," suggested Whitehead with a look of satisfaction on his face, "let's make ourselves comfortable over there under the pepper tree. I know there are questions which you would like to ask, so let's begin."

"Well," said Miles, "you spoke of a 'Scriptural Hour' in connection with the Great Pyramid. How long was this period and what are the dates of its beginning and ending?"

"It's a period known in Scripture as the 'one hour' period. Its purpose was to awaken humanity to start a process of change, expansion, and increased spiritual understanding.

"The 'one hour' period is a sort of 'cusp' between the tired old dispensation and the time of preparation for the new Golden Age about to dawn upon the world. This short time period of fifteen years contained many changes and events, some quite drastic, meant to jolt mankind out of its state of idleness and disinterest and onto the path toward a fully expanded Five-Fold life. The 'one hour' period appears to have ended August 20, 1953. It should be clearly understood that time frames which have been established by human beings are not always the same as those set by the Creator. Scripturally sound prophesies are not to be confused with the predictions and interpretations of mortal man."

"What happens during and after the fifteen-year period?" pursued Miles.

"Determined people wanted to find as many answers as they could," laughed Mr. Whitehead, "and many of them, I suppose, so that they could accumulate great wealth for themselves. Fortunately for the world, nothing so definite is made known. This was to be strictly a preparatory period for mankind to formulate a foundation for entering into the coming Golden Age."

"How much preparation does mankind need?" queried Miles.

"With some exceptions, mankind is in a deplorable state, spiritually, mentally, physically, socially, and financially. If one has any idea of what a perfect human race should be, he can see how much preparation is necessary, and what a tremendous change must take place in the hearts of men and women in order to enter the path to expansion, said Mr. Whitehead.

"The human mind cannot solve problems because the human mind alone wants only itself to survive. However, when the human mind is one with the Mind of The Creator, the two as one can solve any problem, he continued."

"Twenty years ago I knew exactly what a perfect world would be like," said Miles. "Ten years later, I wasn't so certain and now, I'm only sure that I don't know. I imagine, however, it's a world where everyone is honest; where all lawyers are truthful; where all doctors know and practice their profession for the sole purpose of benefitting the patient; where politicians are sincere and capable lawmakers; where everyone has enough good and wholesome food to eat, food, clothing and shelter; where child labor is unknown; where personal liberty, personal safety and security are realities. Do you agree with me, Mr. Whitehead?"

"In most things, yes," answered Mr. Whitehead. "I can see that you've given the matter considerable thought, and have reasoned correctly. However, there are a few discrepancies in some of your opinions. Shall I point them out to you?"

"By all means," said Miles, anxious to learn as much as he could.

"In the new scheme of things," began Mr. Whitehead, "men, as you say, will be honest. As a matter of fact, only people who are truly honest can possibly enter on to the path of true expansion. At this time the world is full of so-called honest people. But are they really honest? While things are going well for them they are honest; but when adversity hits them, their honesty takes a back seat and they quickly resort to dishonest methods of operation to avoid financial inconveniences. A really honest person is honest all the time regardless of circumstances.

"As far as doctors are concerned, when everyone is living decently and ethically, knowing and following natural, universal laws, there should be little or no sickness and therefore no great need for doctors.

"Lawyers and politicians would be practically unknown. When everyone is honest they recognize one another's rights and are always ready to respect them. So, aside from a few simple rules there would be no necessity for laws and lawyers. The golden rule, which represents itself in every major religion and discipline, of doing to others as you would have them do to you, is one of the cardinal rules men and women should follow every day and in every way that they live. Where there are neither laws nor lawyers, there would be no need for politicians, who are the law makers.

"The new plan of progression is at our very doorsteps and is so far advanced over the old one by which we have been living and merely existing that the present laws will have no bearing on it.

"The personal liberty of which you spoke is also more or less dependent upon laws. The more laws, the less personal liberty. Each law brings about more restraint and while it is necessary now to have some law in order to prevent the shrewdly malicious from taking advantage of the ignorant and foolish, as well as those whose disabilities prevent them from the ability to make rational decisions from time to time, modern politicians might better promote personal liberty by bending their efforts to repeal certain laws rather than constantly enacting new and superfluous ones. After all, laws are really made the the law breaker, not the honest people.

"Food, clothing and shelter could have been obtained absurdly easily at any time in the past 6,000 years had it not been for the fact that it would have created a state of slow-going and laziness which would have halted the progress of mankind, causing them to sink into a state of degeneracy and stagnation. Let me give you a striking example of how easy it is for this to take place.

"The last thing which you mentioned was child labor. Here again you have touched a subject which is involved in the changes ahead. In the time to come, there will be no child labor. But then again, there will be no labor of any kind. Perhaps you remember how Adam and

Eve, having been driven out of the Garden of Eden, were forced to live and sustain themselves by the sweat of their brow. They found out quickly just why the Book of Genesis called this a curse. This torment was to continue to the end of the old dispensation. In the last few decades we have already experienced forewarnings of that time. Destiny seems to have eased up a bit and has permitted man to create for himself many labor-saving devices and new technological processes. There is always room for new ideas. However, we won't be living in the new times too long until no physical labor at all will be required to procure all of the necessities of life. I seriously doubted this very thing until finally, when I had earned the right, I was shown how simple this path will be."

After being engrossed in meditation for a short time, Miles asked, "How is it possible for us to make such an incredible change in such a short period of time?"

"In the prophetic interpretation of the Pyramid of Gizeh are turning points which have a distinct bearing on the destiny of mankind. These turns are always to the west. In pyramidal terminology, one who has gained enlightenment and is striving to expand into life is referred to as having 'turned West.'

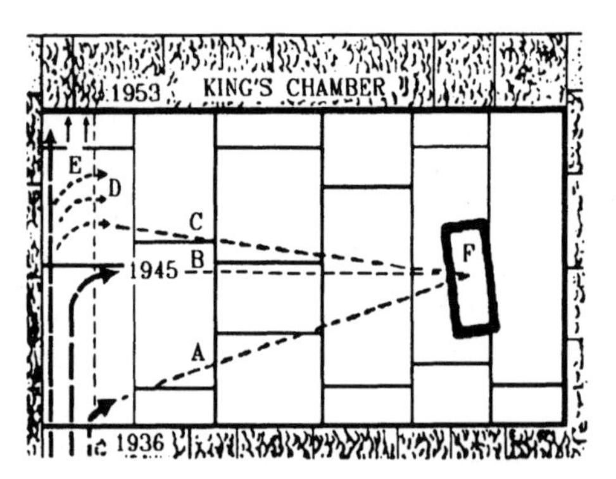

"A", "B", "C", and "D" represent individuals who "turn West" — turn to life — in the King's Chamber between September 16, 1936 and August 20, 1953. "F" is the open Coffer, meaning "victory over death." "E" are the masses that "go South to destruction." They become identified with the "dead end" passageway (3X).

"While there is no direct connection, it is interesting to note that the ancient Egyptians believed that Eternal Paradise lay to the west of the River Nile. As a consequence, pharaohs and fish wives, kings and knaves, all found a final resting place there. The 37 Egyptian pyramids, which were tombs for the ruling classes, were all built on the west side of the Nile. So the idea was commonly accepted among the Egyptians that dying was, in reality, 'going West' to the Eternal Paradise. During World War I, the British soldiers who fought coming up from Egypt to capture Jerusalem adopted the phrase 'gone West' in reference to death, and in turn, passed it on to the American soldiers. Now it is a term used to mean one who has chosen the forward and upward path.

"But while the ancient Egyptians were obliged to die in order to 'turn West', we who are alive at this time must live in order to do so. To 'turn West' now means to enter a new dispensation where good has the edge over evil.

"After the fifteen-year period people can be divided into two classes: those who have already 'turned West' and those who now or some time later will do so", he explained.

Mr. Whitehead took an envelope from his pocket and quickly sketched a diagram of the King's Chamber on it. With this serving as an illustration, he continued,

"In this latter class will be many who through their over-developed viciousness will annihilate themselves. In fact, millions of the world's present population will come to an end because of their refusal to 'turn West'. Remember what I related to you regarding the flu epidemic of 1918, please keep in mind that this or a similar situation could happen at any moment.

"Individuals will begin to separate themselves from the masses and 'turn West', leaving the southern path (the path of destruction) and embarking on the Path of Life and expansion. Once one has made the decision and has 'turned West' there is nothing to fear because turning West means constant improvement in all departments of life. By living the right kind of life today, you are outwitting the destructive forces of tomorrow.

"Already, the forces of good and the forces of evil on the face of the Earth are so equally matched in power and influence on mankind, that man finds himself in a neutral state. Neither of the two powerful influences unduly sway him and he is at liberty to choose either the westward

Path to light, life and liberty or he may choose the southern path to defeat, destruction and death.

"Taking a neutral position or condition at this time is a most critical and serious choice. Man is entirely responsible for the actions he takes and the path he chooses. If he chooses the downward path he does so of his own free will and quite naturally will suffer the consequences. If he decides to choose the Upward Path, then he will find himself marvelously rewarded for his outstanding judgment.

"Great days are ahead, Mr. Miles. In fact, they are already here. It is our job, yours and mine and tens of thousands of others, to enlighten people about the things to come so that they too can outwit the coming evils of tomorrow. Each one who rejects the influences of the evil forces decreases their power and the effectiveness of these destructive legions while at the same time, increasing the power of the White Forces of good by the decisions of man to turn West."

Glancing at his watch Whitehead exclaimed, "Three o' clock! I hadn't realized time was passing so swiftly. I'm afraid our discussion must end now, Mr. Miles. I have an appointment in Casa Del Rey at four o'clock, but I'll see you over here in the valley at about the same time tomorrow."

Chapter 6

Inevitable Events

Once again Whitehead and Miles sat down in the cool "hut" over in the fertile little valley and began to eat their enjoyable lunch. After discussing some points that had not been entirely clear to Miles regarding the time measurements in the Great Pyramid of Gizeh, Whitehead asked,

"Do you remember, Mr. Miles, the morning of January 1st, 1900, when every one greeted each other with 'Happy New Century' instead of the usual 'Happy New Year'?"

"I most certainly do," replied Miles. I remember it as clearly as though it took place yesterday."

"Well," continued Whitehead, "the 'Happy New Century' didn't turnout to be so 'happy' after all. In fact, for the mass-minded it has been anything but a happy one, and I dare say that the last half of this particular century, which began January 1, 1950, wasn't nearly so 'happy' for the mass-minded as the first half.

"Of course, for those who are traveling West, for those who are traveling the Upward Path, none of the acts of destiny intended for the awakening of the mass-minded will affect them, other than for good.

"The mass-minded are, in reality, two separate and distinct groups. The first group is the incorrigible mass-minded that will not 'turn West' under any circumstances. The second group are those who will eventually 'turn West' and will travel 'The Path', but only after much prodding and no small amount of persuasion. They must not be urged beyond their free will. For this reason it will be necessary for the White Forces to acquaint them with adequate educational material so that they will desire to 'turn West' and expand into the five-fold life of their own accord.

"There is a 'deadline', so to speak, for the mass-minded to 'turn West'.

"For the exceedingly large group—those who are not willfully mass-minded—there is considerable time remaining. Some of them will be exceptional cases and may have a longer time to 'turn West'. However, 'incorrigibles' could be eliminated at any time from physical embodiment as soon as they are 'discovered'. This means that millions upon millions will have ceased physical existence during the last half of this century and even into the new one whose dawn is upon us, who might otherwise have gone on living far into the new millennium which began on January 1, 2000 A.D. This elimination will continue on this new century.

"There is another thing you must keep in mind. Throughout the final years of this century new souls came into physical embodiment. Those born during 1949 were fifty years old when the new century made its entrance. Just like those born before the last half of the 20th century, they will have to make up their minds whether or not they are going to 'turn West' and live, or go to an untimely death intended only for the deplorably mass-minded.

"This also is very important. Everything that took place during the last half century is for the express purpose of causing the mass-minded to 'turn West' or, on the other hand, to remove from physical embodiment all those mass-minded ones whom Destiny knew would never 'turn West'.

"Scripture tells us of two great battles that will be fought.

"The first of the two battles is known as the Battle of Jerusalem. It is often called the Battle of Jehosophat due to the fact that it is to be fought partly in the Valley of Jehosophat which is close to the city of Jerusalem.

"Events are already being arranged by Destiny to bring about this battle so keep your eyes closely fixed on this area of the world and the things which are happening there each day as the future continues to unfold.

"On the other hand, the Battle of Armageddon that we hear so much about these days will not be fought until after the Battle of Jerusalem has already taken place.

"The two battles, Battle of Jerusalem and the Battle of Armageddon, are not one and the same battle.

"Amateur theologists and amateur prophets constantly confuse these two battles. No one who does not know the difference between them, or worse still, mistakes them for being one and the same, can possibly be correct concerning other Scriptural prophecies.

"Before we go further into this, let me explain the divisions of time which are coming. It is made up of 'seven weeks' of time. In Scripture, a 'week' is as seven years. Seven 'weeks' could therefore be forty-nine years. Old Testament Law made the fiftieth year a Sabbath—a Sabbatical Year. Thus, in the fifty-year period there are seven separate and distinct periods of seven years each, plus a 'period' of one Sabbatical year, making a total of fifty years. Scripture also tells us of seven Tribulations, sometimes referred to as the seven plagues, which will occur to mankind and the world during these days or the final seven 'weeks' of time. They are as follows:

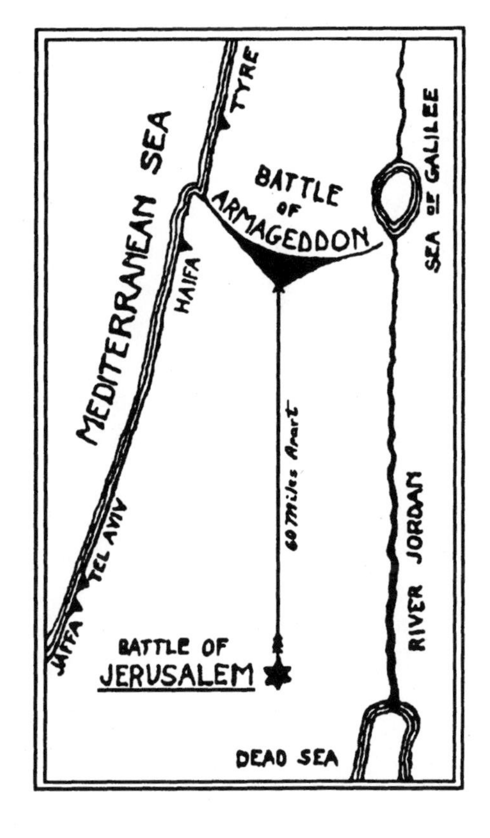

- *The Battle of Jerusalem*
- *Blistering Heat*
- *Destruction of Ships*
- *Water Made Bitter*
- *Sun and Moon Smitten*
- *Men Seek Untimely Death*
- *Battle of Armageddon*

"The above list of the seven Tribulations is the events that will occur during the final seven 'weeks', but not necessarily in this order. The Battle of Jerusalem begins the period of seven Tribulations, and the Battle of Armageddon ends it.

"The Battle of Jerusalem will not be a world war. It will be a localized battle with international repercussions.

"This is what Zachariah of old in chapter fourteen, verses two and three, had to say about it:

"'For I will gather *all nations*[4] against Jerusalem to battle; and the city shall be taken, and the houses rifled, and the women ravished; and half the city shall go forth into captivity, and (the) residue of the people shall not be cut off from the city.

"'Then shall the Lord go forth, and fight against those nations, as when he fought in the days of battle.'

"Remember that at the present time, the Battle of Jerusalem has not yet taken place. This means that the "seven weeks" of time and the "seven tribulations have not yet begun either. There will be no mistaking the true Battle of Jerusalem when it does finally begin.

"After this battle, Jerusalem will be freed from the 'hand of the foreigner' for the first time in many a long century.

"The completion of the Battle of Jerusalem will very definitely indicate to all those who have 'eyes to see' that the mass-minded world is then definitely in the beginning of the end of this miserable, old exact hour of time when this battle will begin and end.

"The seventh and last 'week' is of a pivotal nature. During this period, time all but runs out in the old dispensation. It is during this time span that one third of the Earth's population, the incorrigibly mass-minded of that day, are permanently removed from physical existence. It is also during this 'week' of time that the vast army of 200,000,000 comes down from the 'North Country' into the land of Palestine 'to take a spoil' and the Battle of Armageddon begins to take place.

"The Battle of Armageddon, won't occur until some years after the Battle of Jerusalem, and sixty miles further north. If you remember this, all of the thousands of false prophets in the world will not be able to confuse you regarding things to come.

"It is worth noting that in Hebrew, the word Armageddon means 'Mountain of Megiddo'. Megiddo was the scene of many decisive battles in antiquity and the town became the symbol of the final disastrous rout of the forces of evil.

"The Battle of Jerusalem in the Valley of Jehosophat is small in comparison to the Battle of Armageddon. In the Battle of Jerusalem, there may not be more than two hundred thousand combatants all together. In the Battle of Armageddon, there will be two hundred million.

"In the Battle of Armageddon, the huge 'Army of Gog' will do no fighting. Gog is located and is part of the northern most countries around the Black Sea known as 'the north' and is also known today as China. But the various factions within the "Army of Gog" will viciously fight each other when Destiny brings about the proper psychological moment.

"This will be the greatest slaughter of all time. Only one invader out of six will remain alive and flee from the country. This means that practically one hundred and sixty-six million mass-minded warriors,'stooges and dupes' of the 'King of the North', will be slaughtered in an amazingly short period of time.

"According to Scripture there seems to be a strong indication that during the 'seven weeks' of time, the 'King of the North' and his allies will not be permitted to 'practice the art of war' in any form. For this reason 'the King' comes down to Palestine with a huge, but almost unarmed army, depending instead on vast numbers and sheer force to overrun and take the 'land of unwalled cities'—Palestine.

"This should be remembered: Jerusalem, after the Battle of Jerusalem, will not be the same city it is today. After the battle, Jerusalem will be rebuilt along new, modern and unusual lines. It will then become the Key City of the entire Earth, and continue in this status until after the Battle of Armageddon.

"There will be a period of time when 'all things are made new' for those who have 'turned West'. It will be a most joyous 'seven weeks' for those individuals everywhere.

"For the mass-minded, the future is not at all bright. There will be wars, plagues, pestilence and death, but all of these exceedingly unpleasant things are for their awakening.

"At this moment we are not yet living in the seven 'weeks' of time period. We need to know the things to come so that we will not be alarmed and bewildered when they start taking place in Jerusalem and everywhere else."

Here Mr. Whitehead paused for a moment to check the time. Glancing at his wrist watch he exclaimed, "Five o' clock! I hadn't realized the time was going so fast. I'm afraid our discussion must come to an end, Mr. Miles. This evening I must go up north again and I'll be gone for quite some time. Now I must get back to Casa Del Rey and pack.

"I'll write to you while I'm away. There is further information I want to send you regarding the five departments of life. I will be returning at the beginning of summer and by then you should have absorbed and digested what I have told you already, along with the information I will be sending to you by mail. That way we won't have to waste any time moving on to higher and more profound subjects."

CHAPTER 7

The Five-Fold Life

The days that followed were busy ones for Miles. Life had taken on a new and brighter meaning since Mr. Whitehead had come into his life and the information which his eager mind had absorbed was now beginning to digest. This zest for life sharpened his perception so that he began making discoveries about commonplace things which had gone unnoticed for years. He marveled that he could have been so unaware and foolhardy.

He persisted in his efforts to make play out of work, and succeeded to such an extent that work was no longer drudgery. This in itself was an unbelievable achievement. He had always been a worker and had never shirked a hard or disagreeable task, yet he had always detested it and been resentful of the necessity of having to work.

More and more often he went to the top of the ridge at sundown to watch the "flare up." The knowledge that he was privileged to see this phenomenon gave him confidence in his mental ability, and this confidence, in turn, enabled him to clarify many points which had been more or less obscure. He meditated on his fascinating experiences and Mr. Whitehead's explanations recurred to him on a regular basis. Invariably these remembrances impressed him more vividly than had the actual telling, an indication that his subconscious mind was beginning to function as an asset instead of a liability.

The "one hour" (fifteen-year period), and the seven "weeks" became of great interest to him. Mr. Whitehead had said that those who knew what was to take place and were preparing for it had nothing to worry about, and although Miles knew in part, he was still anxious because he had much to learn and time was slipping by so quickly.

With his newly awakened understanding he read the signs of

times. Minor day-to-day events that were passed over by most people without a thought to him were indicative of an increasingly troubled and chaotic world, rapidly approaching the end times.

In meditating on the conditions of the masses, Miles' chief concern was how they could be helped. In fact, he had always felt such concern, but he had always thought himself to be one of them. Now he suddenly realized that he was being separated from them, that a wedge of individuality was being driven between him and the general rank and file of men and women. The thought almost frightened him. To be separated from the masses? How could he get along without them? How could they get along without him?

But as he thought further on the subject it occurred to him that the masses had gotten along without him for a long time, ever since he had moved to the Pacific Coast. They would somehow continue to get along without him now. As for helping them, it was practically impossible when he was one of them. They were like people in a gigantic whirlpool. Around and around they went, drawn nearer and nearer to the vortex. One man on shore with a lifeline could be a thousand times more helpful than could the best swimmer out amongst them. This idea of being on shore with a lifeline appealed to him, yet he still felt a strong desire to mingle with the masses. His emotions were so muddled that he gave up in confusion. "If only Whitehead were here," he thought.

At long last the letter arrived. He was disappointed at its brevity, but in it Mr. Whitehead told him that, first, he wanted Miles to read carefully the "monograph" which was enclosed; to study closely the information it contained; and to make it as much a part of him as he possibly could. He should know it so thoroughly that each of the five departments of life would stand out clearly whenever he entered a particular one, either in thought or in action. Then, after a reasonable period of study, meditation and understanding, Mr. Whitehead would write again at greater length, reviewing the "monograph" and revealing additional information.

Miles read the information through, and then, with growing interest, read it again carefully. This was exactly the instruction he needed and he had been craving. Now the five departments of life became clear and definite in his mind, and the necessity for being able to mentally picture every major phase of life was plainly brought out. Now he would be able to chart his life by this method and prepare for the great changes the New Dispensation would bring. He studied the "Monograph" intently and it is reprinted here for all to read and discern its message.

A Special "Monograph" Containing Such Information from Exoteric and Esoteric Sources Regarding the Purpose and Value of the Equal, Symmetrical, and Constant Development and Expansion of the Five Departments of Life

Ages ago, Destiny populated this Earth with a race of mortals and placed before them a path upon which they were induced to travel. This path, leading constantly forward and upward, would eventually bring them to a place called Immortality and then, if they chose, they could be one with the Incorruptibles.

However, the path from "a clod to a God" which Destiny had provided for them was a long and exceedingly arduous one; and often Destiny had to scold them to keep the mortals moving onward and upward. Then the day arrived when they began to comprehend that there was a purpose in all this. The further along they progressed, The Path became easier to navigate. As they recognized this, persuasion was less necessary and they more willingly hurried on their way.

Man has come a long way since his meager beginning long ages ago. He is about to reap a vast reward for all the pain, sorrow and anguish the journey has caused him.

Throughout the lifetime of man upon this Earth, his actions and thoughts fall into five separate and distinct divisions, or departments. His success in living depends entirely upon how well he has expanded and used these five departments of life. A star aptly represents them,

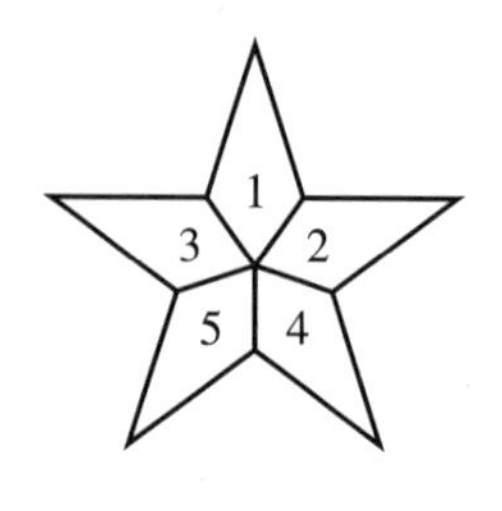

with each point indicating a particular department. The uppermost point represents the Spiritual; the upper right one, the Mental; the upper left, the Physical; the lower right one is the Social; and the lower left, the Financial. These five divisions completely cover every phase of an individual's life, regardless of what diverse names they bear. They remain unalterable, constant and resistantly fixed. And understanding each one separately and in combination with the others is actually a very simple process.

Life holds but one purpose. A purpose which is inexorably prescribed by Destiny, that each human being, starting from the center of his individual star, must fill out the departments of life represented by the points, evenly and symmetrically, taking care that one point not be greatly developed beyond the other, but rather that they be expanded equally and evenly.

Herewith are five stars. Each one shows a particular department of life which is overdeveloped at the expense of the other four. Lacking the knowledge that there are five separate and distinct departments, it is impossible to develop, enlarge and expand all of the five evenly and symmetrically. At least one is always bound to be eccentric and abnormal. Is it any wonder, then, that the average person experiences so much sorrow, ill health, fear and poverty?

The first star of the series shows a person who is overdeveloped spiritually or religiously. He is "all heart and no head." Except along very narrow and bigoted lines, reason, will and judgment are warped and stunted. Although called a spiritual or religious type, very often this person's beliefs are so narrow and intolerant that he really isn't spiritual at all, merely fanatical. His friends are of like character, and because he imagines them to be more religious than himself, he becomes jealous of their activities and is usually antisocial. Physically, this person is only a fraction of what he should be. He is inclined to dyspepsia, anemia and nervous disorders. He possesses little of this world's goods. Not that he wouldn't accept what was offered with the eagerness of a miser, but because he is in such an appalling mental state and is so judgmental, straight-laced and unyielding in his religious beliefs and practices, he drives everything of an abundant and opulent nature from him.

Spiritually overdeveloped

The second star represents a person who is overdeveloped in the mental department of life. He lacks spirituality, reasoning that if God cannot be found between the covers of a book or in a test tube then most certainly there is no God. Like the spiritually overdeveloped, he is narrow, bigoted and intolerant of those who do not share his opinions. Caring little for physical exertion, his health is usually in a deplorable state, while socially, he confines himself to associates of the same mental inclination as himself. Living constantly in theory, he is highly impractical and only by the greatest effort can he manage to have enough finances to support himself comfortably.

Mentally overdeveloped

Physically overdeveloped

The physically overdeveloped is shown by the third star. Strong and robust, he is the typical male, noisy and domineering. The spiritual side of his nature has never developed; but much can be said to his credit. He seldom denies the existence of that which he doesn't understand or in which he has little interest. He is below average mentally, running more to muscle than to mind. Socially, he is popular with that class of people who see beauty in the movements of bulging muscles. He is usually in modest financial circumstances due to the fact that through physical effort he is capable of earning enough money to satisfy the wants of his physical nature so he does not concern himself about accumulating wealth until his earning capacity begins to fail him as he ages and begins to lose his edge.

Socially overdeveloped

Financially overdeveloped

The fourth star typifies the individual who is overdeveloped in the social department of life. He is of the hail-fellow-well-met variety. An innocuous handshaker who, due to his affable nature, comes by many free meals and alcoholic drinks. Spiritually, mentally, physically and financially he is in pretty poor shape, having neglected them all in favor of being sociable. Occasionally he falls into a remunerative political job or is retained as a professional greeter, but as a general rule he leads a hand-to-mouth existence. Of course, there are exceptions to this type of individual. Those who involve themselves in the political scene are many times able to bring to themselves wealth beyond their expectations. These are those who work in the higher levels of politics, many times for a specific political party or person. And it is through their experiences in this capacity that they are able to capitalize on their familiarity and "inside" knowledge to write books and participate in speaking engagements which can and many times do bring them great amounts of wealth in a very short time. But this too is transitory. Man's attention span is so brief that the publicity these individuals attain is short-lived and without the knowledge and inclination to invest these earnings wisely, they will find themselves again in a position of having to find a new place for themselves in society.

The financial department of life is shown overdeveloped by the fifth star. This type of person might very reasonably and easily have been an all-around, evenly-developed individual before the mania for money struck him. He has the ability to fill out the other four points of his star, but once having succumbed to the craving for wealth he rapidly becomes stunted and dwarfed in everything else. He goes to his grave with an insatiable desire for more and still more wealth, and fears and abhors death because it deprives him of his material gain. Usually he is devoid of friends and has sacrificed his spiritual heritage for his lust for money. His health is rarely good because of neglect and his mentality is limited in scope to schemes for more wealth and power in spite of his usually above-average intelligence level.

These five stars show the extremes in overdevelopment of one particular department of life. Seldom does any one person exhibit such an exceptionally eccentric type. Usually it is a variation or combination of these types. Often two points are emphasized; in some three, and rarely, four are well-developed. However, there is always one point which is woefully neglected and which retards progress in the other four. Destiny dare not permit an individual to become too highly developed in four departments of life without developing the fifth. Such a being would become a colossal menace to society, especially if the spiritual department was the dwarfed point of his star.

Now we come to the star which indicates an even, equal and steady development in all five departments of life. This person shows little promise at first of being in any way exceptional. But as he fills out his star with an attitude toward balanced living he becomes powerful and power-filled. He starts accomplishing things, and his accomplishments are real and lasting because he has built them on a symmetrical foundation which is solid and strong. Results are noticeable to everyone when this individual has even half filled-in all the points. From then on he will completely fill-in the entire star in a very short period time and become a competent, all-around super individual. He is in harmony with all that is constructive in both the visible and unseen worlds. Forces which would frighten the ordinary, commonly-developed person into convulsions are his friends and allies. (Read more about "Strange Forces" later in this book.) They race to do his bidding. But this is not the end.

There is no end. Once a star has been completely filled in there are still greater things one can seek and make every effort to attain. These star points are capable of unlimited extension and can continually be pushed out into added achievements in all five departments of life. This process of extension is

shown by the last star with the elongated points. There is no limit to the length of these points. Long points become longer ones as the individual marches onward in his conquests. And when one has experienced the thrilling expansion of all five departments of life, there is no turning back. From then on it's a search for "more worlds to conquer."

To him who by this humble discourse, and its still more humble illustrations, manages to "catch the fire" in his quest for expansion in the five departments of life, there are no limits, no sorrows, no more darkness. From then on, all is bright, joyous, radiant and thrilling.

CHAPTER 8

Observations Concerning the Brain

Previous to the time when Mr. Whitehead had sent him the monograph on the five departments of life, Miles had labored under the delusion that life was an intricate and complex problem, and that its different phases were so closely intermingled that no one could possibly separate them into parts. But now, all that had changed. Miles realized that life was not complicated at all; that it was in fact exceedingly simple. Life no longer baffled him. He felt certain that he could cope with it and master it and the thought caused a pleasant sensation of strength to permeate his entire being. In a few days, true to his promise, Mr. Whitehead wrote, and after a few personal items, said,

"At the present time your spiritual department is beginning to come into its own. If you refer to the monograph which I sent to you, take note of that part dealing with the spiritually overdeveloped, noting that there is a great difference between a religious person and a spiritual one.

"For centuries, mankind has been taught that 'God is in His Heaven.' And although it has been intimated that 'God is Within,' it seems to be completely ignored, perhaps because people do not understand its significance and so they continue to seek God in some vague, distant place in the starry firmament. These people are merely religious. Their God is always without, never within. The results of their efforts are hundreds of dissatisfied sects, institutions and organizations, each claiming to know the one and only path to Heaven. You must continue to develop your faith, remembering that faith is the substance of things hoped for, the evidence of things not seen.

"In contrast to this, those who learn and know, and who practice what they know, who have 'turned West' in all five departments of life, are beginning to experience something new taking place within them. They are feeling the first effects of true spirituality.

"There is a vast difference between memory and understanding. The average university graduate has crammed his head with a huge amount of information (and no small amount of misinformation) which he has memorized but it's of little value to him because lacking understanding, he doesn't know how to apply it. Knowledge and facts are of no use unless they are applied to one's life; and no individual who has not developed an ability to reason and to judge in advance of learning can put such training to use. As a consequence, he is of little value to himself or to the world.

"The mental department, like the spiritual, is capable of being grossly misunderstood. Many people confuse intelligence with education. The mental department is divided into three realms: the conscious, the subconscious and the superconscious.

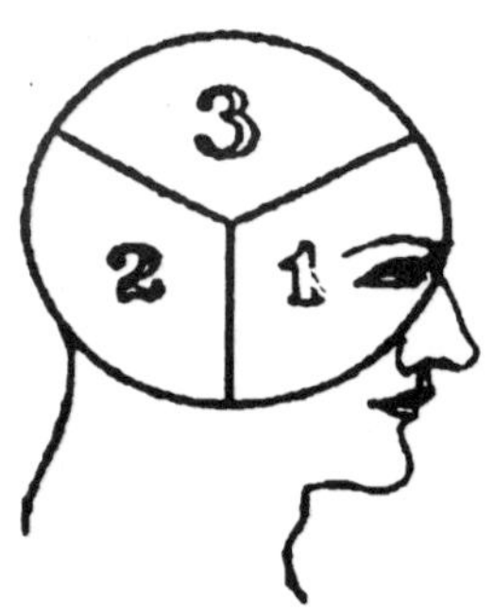

1. **Conscious realm of mind**
2. **Sub-conscious realm**
3. **Super-conscious realm**

"The subconscious takes care of habitual thoughts and actions and consists of memory, imagination, beliefs, affection, and emotion. The superconscious is the realm of the spiritual side of man's nature and works through the subconscious, using the qualities of inspiration, intuition and genius. But about these two realms we are not directly concerned at this time. We are, however, directly concerned with the conscious mind, for here dwell reason, will and judgment. The development of these three mental qualities marks the difference between education and intelligence. If these qualities are lacking, education is merely a matter of the ability to recall information. Here you must get the Mind of The Creator working in concert with your human mind. Plug into the source of all knowledge with effectiveness and release this gift to the mass-minded ones.

"An educated man with a highly developed ability to reason and judge and a strong will is, of course, invaluable. You seldom find him working for someone else. Then there are comparatively uneducated people who have a more highly developed reason, will and judgment than do the college-trained ones and are usually found at the head of a great business undertaking or widespread public projects. So regardless of how limited one's education may be, if he has reason, will and judgment; if he knows what he wants to do; if he has sufficient zeal, fervor and enthusiasm, he will succeed in doing great things in spite of any handicaps including age, environment and circumstance. He will outwit tomorrow.

"In the physical department of life there are two causes for most afflictions. One is psychological, the other physiological. When I first met you on the ridge above your home, Mr. Miles, you were suffering from rheumatism and stiff joints. Now you never mention it because it has disappeared. It disappeared because it was of a physiological origin. You were extremely critical. You criticized your wife, your children, your associates and even yourself. But when you changed your thinking, your critical attitude disappeared and along with it your rheumatism and joint pain. I am attaching herewith a short list of physical afflictions with a psychological cause. It was given to me by a noted psychoanalyst but is in no way a medical diagnosis or representative of a medical treatment. Only licensed physicians can perform these functions. This is given to you for your own illumination and to ponder what the connection is between mind and body.

Physiological Afflictions and their Psychological Origin

Of the thousands of physical afflictions from which a person may suffer, there are but two causes. The first is of a strictly physiological origin; the second is of a psychological origin. Of the two, the psychological is said to be by far the more prevalent and is generally due to an exceedingly high amount of stress in a person's daily life.

Here is a short list of physical afflictions and their psychological causes. As you may note, many of these afflictions have more than one cause. This is not to be taken as medical advice or diagnoses and does not replace a personal visit with a licensed physician. It is provided for observational use only.

Apoplexy — Brought on by anger, hate or extreme passion.

Back lameness — Burden-bearing thoughts.

Biliousness — Revengeful, traitorous, mutinous thoughts.

Boils and other eruptions — Irritability, impatience.

Baldness — Incompetence, inability, self-consciousness.

Catarrh — Disgust, disdain and false superiority.

Cancer — Dissatisfied love nature, selfishness, frustration.

Colds — Depressions, despondency, "the blues."

Constipation — Nervous tension, worry, lack of poise.

Croup — Intense irritation and confusion.

Deafness — Unwillingness to listen, judge and accept.

Heart trouble — Selfishness, fear, worry, tension.

Hemorrhoids — Prolonged anxiety, fear and worry.

Hysteria — Repression, mental conflicts, selfishness.

Kidney trouble — Inferiority complex, fear of detection.

Liver trouble — Inaction, depression, repression.

Nausea — Rejection of facts or truth, emotional conflicts.

Paralysis — Thwarted or inhibited desires.

Pneumonia — Overwhelming disappointment of long duration.

Rheumatism — Faultfinding, criticism, nagging.

Sore throat — Unconscious resistance to truth.

Spinal trouble — Remote fear of death and eternal punishment.

Stomach trouble — Over-sensitiveness, rejection of facts.

Frigidity — (In either sex) Repression, conflicts, shocks.

Tuberculosis — Lack of freedom or a shut-in complex.

Urinary trouble — Inefficiency, inability, and "I can't."

"Keep this list in mind as you continue your development. Guard your emotions against any interference which may be the source of one of these problems. Stress is the greatest cause of physical as well as mental ailments. Make every attempt to eliminate it from your life and prepare yourself to teach others how to do so. Also remember that not all afflictions are psychologically based. There are many which are truly physiological and you must learn to be able to tell the difference and thereby seek the attention of a medical doctor when an appropriate situation dictates. Be careful not to give medical advice or opinions to anyone, especially someone you do not know. Apply these truths to yourself first and foremost. Then, if asked by someone for your feelings on the subject, only at that point in time should you even bring up the subject of possible psychological causes for some infirmities."

CHAPTER 9

The Stone Which The Builders Rejected

A few days later, Miles had another letter from Mr. Whitehead concerning more about events as foretold by the Great Pyramids of Gizeh.

"We have already traced the succession of events as foretold by the Great Pyramid of Gizeh, from the time of Adam up to the ending of the fifteen-year period on August 20th, 1953, and the 'One Hour Period.'

"Now let us examine more closely the King's Chamber. We have seen that the south wall marks the ending of the 'One Hour Period,' on August 20th, 1953. We have run up against a blank wall, and there is nowhere else to go but West, or UP!

"We know that August 20th, 1953, has come and gone and the mass-minded are still very much in evidence. What has happened? A logical and unhysterical view of the whole thing clearly indicates that humanity, in its forward movement, has reached a point where further material 'evolution' ceased. Only those individuals who have chosen to 'turn West' onto the Path of Light have any chance to continue forward—to expand—as you will remember from the little diagram I drew of the floor plan of the King's Chamber.

"In moving through the second low passageway, that is prior to September 16, 1936, humanity had become accustomed to being bent over with eyes to the floor due to the low ceiling but upon emergence into the King's Chamber with its 19 foot ceiling, the majority continued in this cramped position not realizing there was anything more than just the floor under their feet. They had limited vision considering these circumstances and became 'stuck' due the limitation of their point of focus and their refusal to grasp the new position in which they found themselves. They simply did not take the time or have the inclination to look around them and see other possibilities.

"As you will remember, individuals had the opportunity at any time to 'turn West' toward the Wall of Light and many, many of them have done just that. This turning has given them a longer time in their travel and to realize that they could stand erect; that there was more to envision than just the floor. Here is the beginning of expansion into life for numerous people!

"What about the great mass of humanity that did not 'turn West' during this period? As we know, many millions of them have been removed from physical embodiment never having known they could, at any time, have chosen the Path of Life. Then, too, there are those who reached the South Wall and stopped there, trying in vain to move on but getting nowhere. These are the ones the White Forces try to work with. All they can do is attempt to gain their attention long enough so they will look 'West' and realize there is somewhere else to go. Then perhaps they will choose the Westward Path to light, life and liberty.

"As I said before, mankind's growth in a material way ceased. What then of those who 'turned West' and escaped this? Their growth, or 'expansion', has continued, but in a Spiritual way. They soon realized they could stand erect and look around. There in front of them was the 'open treasure chest'—the symbol of the Resurrection; the new birth. Here indeed is the greatest opportunity which has ever presented itself to mankind. We are living now in this time and it is up to each one of us when and how quickly we will take advantage of this gift.

"As we look at this 'open treasure chest' we realize the top is off and open to that which is above. So naturally our gaze follows up. Here is the way—the Upward Path! As we look upward we must realize that, as our growth in the material way has stopped and our growth or 'evolving' is now Spiritual, we are no longer Earthbound by material laws and our movement is now upward. Whenever I think of this, the words of the Great Teacher come to me, 'And I, if I be lifted up, will draw all men to me.'

"If you will notice in the diagram here, the construction above the King's Chamber is rather peculiar. There are five slabs of granite, one on top of another, with space between. Also note that the bottom surface of each is smooth while the top is uneven. This indicates that the individual moving upward 'through' these five areas (which of course represent the five departments of life: Physical, Social, Financial, Mental and Spiritual, all solidly established and developed) determines his own rate of travel by his own application of the knowledge he has learned and applied to his own life. In other words, the time indicted by the distance from the floor of the King's Chamber to the pointed roof is flexible, depending entirely on the individual and his own rate of expansion.

"Directly above these five slabs of granite is the pointed roof, which is perhaps of greater significance than any other part of the construction of the Great Pyramid.

"It is toward this point that all expansion, or growth, of the individual moves. This is the apex of the 'comprehension' of mankind.

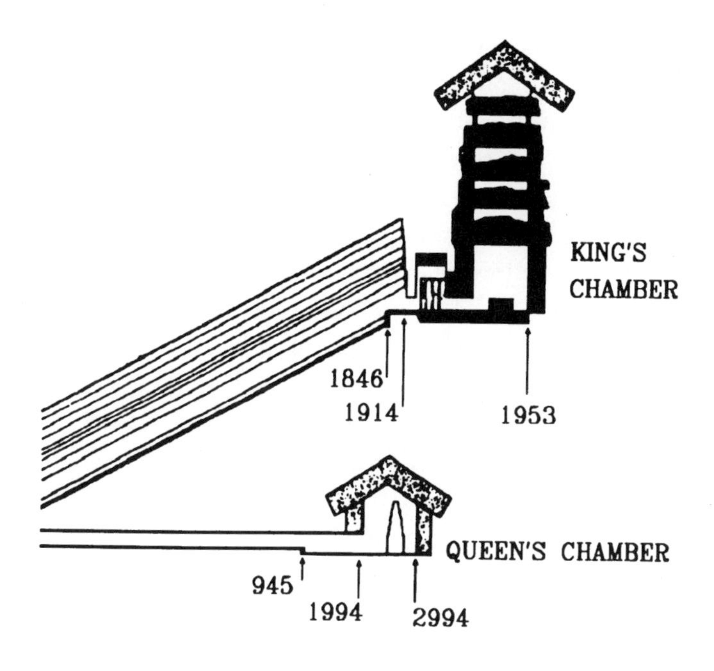

When the individual has reached this point he has successfully passed 'through' and has made a part of himself the solidly-developed five aspects of all mortal life. From here on life points heavenward toward the Infinite and the individual who has attained this position becomes one with the Immortals.

"You will notice that an identical construction is over the Queen's Chamber. The Queen's Chamber is the representation of the new millennium when all that is evil is dispelled from the face of the Earth. No one knows the day or the hour, not even the Heavenly Host, that this time will begin. There have been estimates by many so-called 'prophets' and only time will tell the truth of this matter. Remember what I told you about prayer changing things.

"This exact reproduction in the construction of the roof over the King's Chamber would seem to indicate that the millennium will start for the individual when he reaches this point in his travel on the Upward Path. In other words, the beginning of the millennial period can start for you just as soon as you have successfully begun expansion of all five departments of your life, and this regardless of the calendar date. This can become your own personal 'shortening of time'.

"Now one final thought before I close this rather long letter. As you are no doubt aware, and it can be seen from any photograph of the Great Pyramid, the cap stone on the top is missing. If you will notice in your diagram the pointed roof over the Queen's Chamber is in direct line with the top of the Great Pyramid. Therefore, we can say that Destiny's ultimate goal for man which is symbolized by the roof over the King's Chamber and also over the Queen's Chamber is a clear representation in stone of the "Kingdom of Heaven' which is found not atop the grand structure, but is 'within'. It is 'The Stone Which The Builders Rejected' —the symbol of the Mystic Christ.

"In a few days, Mr. Miles, I'll send you another monograph called 'The Secret of Silence'. I'm sure that it will be of considerable aid in developing your opportunities to the fullest."

A Monograph Concerning The Secret of Silence As Gleaned From the Sacred Archives of the Great Silence Especially Written for All Those Who Desire To Dare, To Do and To Achieve

An ancient adage says, "To tell a thing is identical to doing it." Another that, "Great talkers are little doers." And a more modern one, "A barking dog never bites." All of this demonstrates, "He who talks about what he is doing or what he plans to do, accomplishes but a fraction of what he could do if he kept silent." To keep silent about one's plans and activities comes very near to being the first secret of success in any and all departments of life.

Desire is a peculiar force. It is actually dissipated through idle or enthusiastic conversation about the thing desired. The more one talks about his aims and plans, the less desire he has for carrying through with them. This conversational dissipation is like steam under pressure in a locomotive boiler which is allowed to escape by blowing the whistle when it could have been used to move the train. To dare, to do and to keep silent; but the greatest of these is to keep silent. Without observing this, no mortal being has ever attained anything worthwhile.

While absolute silence brings its own reward under the conditions just set forth, it does not follow that one should keep silent about everything, as is the custom of certain austere religious orders who have taken their vow of silence in order to be in constant prayer and communication with The Creator and for the purpose of the betterment of mankind.

Conversation about anything other than your aims and plans is very beneficial. The small amount of energy dissipated in this case is amply rewarded by an improvement in the mental and social departments of life. Interesting and timely conversation is very necessary to one who is expanding in all five departments of life. So by all means, he should learn to talk fluently, brilliantly, wittily, seriously. He should learn to ask questions, many of them because there is so much to learn. But about the things he desires to accomplish, the ends he attempts to achieve, the things he seeks to do well, tell no one, not even a brother or the closest of relatives. Regardless of whom you tell, desire power will be dissipated and wasted by talking too much about this area of your endeavors.

Writing about your aims, plans or desires has the same disastrous result. In fact, because it involves more muscular action as well as mental ability, writing is even more destructive to desire power than is talking. It makes no difference whether you write to a friend or whether you just write for the sake of writing and show no one what you have written, the effect is the same. In writing, as in talking, it is the releasing of the power of your own personal desires for your life which is detrimental.

In another respect it may be well to mention now that this same dissipation is sometimes helpful. When one is worried, angry, fearful or distressed by continuing bothersome thoughts he or she is unable to properly perform their work. One is nervous, irritable and thoroughly miserable. These are the things you should talk about deliberately to a sympathetic listener. If it is inconvenient or embarrassing to talk about them, then write them down, spending no more than fifteen minutes a day in the practice. Having written it, immediately destroy it without reading it again. That would put the worries and troublesome thoughts and ideas right back in your mind again.

The fifteen minutes of this writing-out exercise should be done at a time when the writer can be free from interruptions. It should be done rapidly and with no thought as to correct spelling or grammar. Every thought that wishes to express itself, no matter how foul, malicious, hateful, nasty or obscene it may be, should be allowed to come forth when writing.

Take every precaution not to so much as mention the things you wish to accomplish. To do so would allow your good desires to be dissipated and allowed to escape along with the evil ones you wish to destroy, eliminate and eradicate. Never write longer than fifteen minutes in any one day. To write longer than that would eliminate so much of the unwanted desire or complex that you would experience very harmful results. There will be a temptation to do this very thing once you have gotten into the habit of writing out these unwanted desires, but you must guard against it. You must firmly make up your mind that under no circumstances will you run over the fifteen minute limit. While it would be pleasant to rid oneself of an unwanted desire or thought or emotion completely in one day, it is wiser to spread the process out over a number of days and eradicate any risk of disastrous aftereffects.

The writing-out practice is of tremendous value when the "vacuum" created by the departing, negative thought-form is immediately filled in with positive affirmations—thoughts that are just the opposite of those written-out.

To rid one's mind of annoying desires, fears, worries, emotions and complexes is of the greatest possible value. To keep silent about the great things you expect and desire to accomplish is a priceless gift. Remember the admonition, "Dare, Do and Keep Silent."

CHAPTER 10

Intellectual Power

After a few days during which Miles contemplated "The Secret of Silence" monograph which had been sent to him, Mr. Whitehead again wrote at length.

"It should be remembered," he began, "that desire is a blind force. In 'The Secret of Silence' it is compared to steam under pressure in a locomotive boiler. To continue this comparison, the steam itself does not care how it escapes, but the engineer does. He would open the throttle, let the steam into the cylinder and move the locomotive and train. No sensible engineer would hold down the whistle cord and allow the steam to escape unless it was for a very definite purpose. Thus you can compare yourself with the engineer, and your desire power with the steam. You must utilize every ounce of 'steam' in your efforts to progress.

"As you know, the safety valve of a steam boiler is for the purpose of reducing excessive pressure by allowing steam to escape. A human being has many boilers, desire boilers, and on occasion one will work up an excessive pressure. Fortunately, each one of us also has a safety valve, so that instead of the boiler being allowed to blow up, the steam can be quickly and safely dissipated.

"Before one has become thoroughly adept in all five departments of life, desire power has a tendency at times to build up pressure behind some troubling inclination which reason and judgment tell him would lead to disaster if it were realized. This is where the safety valve comes into play. Instead of keeping silent about these low-minded impulses, as you would do with good, constructive desires or with aims and plans, you should talk about them to anyone who will listen: friends, relatives, strangers, anybody, just so you talk about them long and often. You should tell exactly what you would like to do if you were to carry out those desires to the letter. It is amazing how quickly these yearnings will leave.

"Sometimes one has desires which you would hesitate to tell anyone. In this case there is another safety valve. Instead of telling these wishes to someone they should be written out. Use pen and ink, and write every word which comes into your mind concerning them. In a few days you will find that they literally flow out through your arm. Let them come in exactly the words which occur to you. Never try for better phrasing. Sometimes the words may be vulgar or obscene; sometimes the same phrase will come again and again. Restrain nothing; let them come. Write for fifteen minutes daily and only fifteen minutes, and as soon as you finish destroy what you have written so that neither you nor anyone else can read it. To read it again would be to put back into the subconscious that which you have just released. For anyone else to read it would be to run the risk of their forming false and erroneous opinions about you.

"This writing and talking-out method can be put to other uses than to eliminate stray and poisonous yearnings. A young man from the East coast wanted to be a civil engineer, an aviator and a radio entertainer, all at one and the same time. He was well-fitted to succeed in any one of them, but his desire vacillated constantly from one to another. Realizing that he was getting nowhere, he asked my advice. I questioned him about the three occupations and soon had him talking at a great rate concerning each of them. When he would run down, I would start him up again with more questions, but each time he proceeded with less and less enthusiasm. Finally, he had dissipated so much desire, he just couldn't go on. They finally seemed so ordinary and commonplace that he lost interest in every one of them.

"Then I appealed to his reason and judgment, unhampered by emotion. He soon decided that a career as a civil engineer would interfere with his hopes for marriage and children because it necessitated traveling all over for indefinite stays, so that was out. Aviation, as we soon agreed, would be tiresome and monotonous once the novelty wore off not to mention this occupation putting him in constantly dangerous positions while flying. So that, too, was out. This left radio work and it took only a short time for him to decide that in this field the constant change of faces and activities would prove so vital and absorbing that he would never lose interest, while at the same time his future family would never have to worry about him and his safety. In short, very calmly and reasonably he decided to confine his efforts to radio work.

"I then informed him that the hunger for all three occupations would manifest itself again in a few days, and when it did he was to keep absolutely silent about the radio work, but was to tell everyone how much he wanted to be a civil engineer or an aviator. He was to 'spread it on thick,' and keep on talking about them until he, himself, was tired of hearing it. In this way he could prevent the desire for these two occupations from ever getting beyond the embryonic stage. To supplement the conversation he was to write about them every day for fifteen minutes.

"Having confidence in me, he promised to follow my instructions to the letter. Several months later I again visited his city and learned that he was employed by the area's largest broadcasting station and was quickly becoming a success in radio work. I visited him at the studio and when he saw me, he rushed up and shook my hand, and said,

" 'Mr. Whitehead, I did exactly what you said, I talked civil engineering to death. I talked aviation to death, In fact, I nearly talked several of my friends to death. But it was worth it. This radio work is the most intensely interesting and'

" 'Hold on,' I interrupted him, 'you are about to talk your radio work to death, too. I've already heard of your success so you don't have to tell me anything about it.'

"That young man, Mr. Miles, will be a leader in his field in a very short time.

"Worries, negative emotions and troubles can be dissipated in exactly the same manner. In this instance, it is much better to write them out than to talk them out. People quickly tire of hearing about other people's problems. If you do talk them out, do so in an interesting and highly amusing manner. Get your friends to laugh with you. Amusing them will hold their attention, while at the same time you can rid yourself of undesirable notions without anyone even suspecting that is what you are doing.

"A man whom I know was afflicted with fear, timidity and bashfulness. Even though he was over fifty years of age, he was more shy and backward than when he was a boy of fifteen. All his life he had been bothered by these three damnable complexes.

"Hearing of the talking-out method of ridding oneself of unwanted fixations, he decided to try it out for himself. For a while, he tried talking to his friends, but it soon became apparent that they were becoming bored to distraction as he continued to tell his troubles in a sad and doleful manner, so he did just the opposite. He talked about them in a lighthearted and amusing way. He made great fun of his fear. He told funny stories about his timidity and was so entertaining that people wanted to listen and begged him to continue. The way he told of different incidents, brought about through his bashfulness, made his listener double up in hysterical laughter. In time these three phobias, fear, timidity and bashfulness, were talked to death.

"Needless to say, they are with him no longer. He has performed exceedingly well in his business enterprises and has accumulated more wealth in a few short months than he had in the previous fifty years.

"When one has learned to control his desires, the next step is to control the emotions, and then the entire mind, beginning with the subconscious. In that realm of the mind, everything we have ever heard or read or experienced is safely stored waiting to be recalled and utilized. Once a person has become zealous about self-improvement and comes upon a problem for which there seems to be no solution in the world, then the thing to do is to take the problem to the subconscious mind. Just before going to sleep at night focus the problem to be solved clearly in your mind. Then go to sleep with the full expectation that you will wake up in the morning with the solution right there for you. During the night the subconscious will sort through its accumulated knowledge and when you awaken, almost always the answer is in your conscious mind.

"For the present, however, there are so many problems which you can solve by ordinary means and meditation that you don't need to concentrate on the subconscious. In time, when you need it, you will find that it has brought answers to questions you never consciously placed with it. Then you can deliberately and purposefully start using it.

"After you are working in perfect accord with the subconscious, a time will come when you begin to contact the superconscious realm of the mind. Through this higher mind you will receive the most amazing insights. As in the subconscious realm, you awaken in the morning with a problem solved and you feel that during the night, you have been to a treasury of knowledge where you received the answers you sought. After a period of time you will consciously remember the source of the information.

"But all these things come in proper order and in good time, Mr. Miles. If you are a faithful custodian of small things, you will be made a ruler over large ones. There is much work ahead of you. It would frighten you to part the veil and see too much right now and become alarmed by the things you would see and not understand. A caterpillar clings to a limb desperately in fear of falling. But when it turns into a butterfly neither height nor depth hold any terror for her and she soars with perfect safety far above the trees in the golden sunlight. Right now, you too are in the caterpillar stage, but just about to burst open and take flight because of your dedication to learning.

"I have purposely written you at length and in detail. This information will be needed by you in the immediate future. You will soon be working among the masses, Mr. Miles, and it is wise to be thoroughly acquainted with certain fundamental principles as well as specific knowledge of certain subjects. It won't be long and you will be faced with an onslaught of many questions and you must know the answers so that you can teach the multitudes of mass-minded who have decided to 'turn West' and who will look to you as their teacher. Do not become weary in your heart. Walk, watch and listen. Dust off your faults that have been hiding in the recesses of your heart, mend your ways and be an example to the mass-minded. Do not refuse a kindness to anyone who begs for it while keeping in mind the adage never do for anyone what they can do for themselves. This is where your development and use of discernment will come into play.

"Solomon wrote that 'he walks secure whose ways are honorable; the lips of a virtuous man nourishes a multitude. He may count on life and his desires will end in happiness. His plans are honest. Life lies along the path of virtue. The fruits of virtue grow a tree of life; hard work always yields its profit. In the hearts of discerning men wisdom makes her home. Love virtue.' (Virtue is defined as conformity to a standard of right; active powers to accomplish a given effect; courage. Webster's College Dictionary, 1995) 'Virtue teaches justice, (strength of mind that enables a person to meet danger or bear pain or adversity with courage. Webster's College Dictionary, 1995), temperance (moderation in indulgences) and prudence (foresighted, sensible, discreet).

"Solomon continues to say that 'wisdom is a spirit, a friend to man; that malice in the minds and hearts of the [mass-minded] prevents them from knowing these hidden things of the Universal Spirit. Seek wisdom in all things'.

"Mr. Miles, what you learn without self-interest, pass along to those mass-minded ones without hesitation or reserve. Always be mindful that it is the White Forces whose wisdom has been demonstrated time and time again with whom you listen, learn and work. And by walking and working with honor and virtue, you will continue to receive that which you desire and which you are instructed to give to others."

CHAPTER 11

Take Control Now

Vacation time had arrived at last. Books were put away and classroom doors closed behind departing students. Students hurried home by various means of transportation. Michael Miles planned to spend a few days with his family and then seek employment as soon as possible. He had only one more year to complete his degree at the University and was determined to see it through to completion.

He sensed a change as soon as he entered the front door of his home. His family didn't seem quite the same. His father seemed to be enthused by a new spirit. He remembered him as a somewhat cross and soured old man, extremely critical and prone to complain on every occasion that presented itself. Now he was actually cheerful! His outlook on life had grown youthful and he seemed at least twenty years younger in actions and appearance. Michael had never actually disliked his father but he had more or less avoided him. After only two or three days at home, to Michael's surprise and delight, they had grown inseparable.

Miles invited Michael to go over to the ranch with him, and Michael accepted gladly. He was so anxious to be with his father who had become a constant source of surprise to him. And he didn't quite know why. While climbing over the ridge, Miles mentioned the Great Pyramid and its prophecies. He took a secret delight in letting Michael know that he was quite knowledgeable on the subject now.

Michael, amazed, said, "You seem to have it all at your finger tips, Dad. Have you told this to any of your friends?"

"No," replied Miles, "but I have explained it to several new-comers at the hotel in Casa Del Rey and they were all extremely interested. I am waiting for Mr. Whitehead to come back so that he can explain how to choose those to whom I should talk."

The Miles' reached the ranch and started to work in the berry patch. The conversation continued along very constructive lines and Michael was amazed at how his father was making play out of work and he even commented on it. Miles gave all the credit to Mr. Whitehead for the changed attitude.

"This fellow, Whitehead, must be a remarkable man," commented Michael, "to have such a beneficial influence upon you. He most certainly must be all right."

"He knows more than any other person I have ever met," stated Miles, earnestly. "And it is practical knowledge, Michael, information that you can use immediately and get immediate results."

"I see that I have a very ardent supporter," sounded a pleasant voice.

Startled, Michael and his father spun around and found themselves looking squarely into the face of Mr. Whitehead, who had approached unobserved. The meeting was a quite friendly one. It was as if the three had been old friends meeting after a long absence from each other. They worked and chatted. Mr. Whitehead related some of the highlights of his trip, Michael related his college experiences and Miles listened and commented. At noon they stopped for lunch. Mr. Whitehead had stopped at the Miles residence to leave his car and Mrs. Miles had provided him with a lunch for all three of them to eat at the "Hut."

They ate leisurely and the conversation soon turned to more serious subjects. A number of points previously discussed by Mr. Whitehead and Miles were explained to Michael. They settled themselves in the living room as the talk narrowed to the subject of the Great Pyramid.

"Your sketch of the Great Pyramid is very good, Michael, said Whitehead. "Your father tells me that you drew it quite some time ago. Where did you get your information about it?"

"It all started with a short paragraph I read in a book," replied Michael, "but strangely and for no apparent reason, it interested me intensely. From then on it seemed that the information about the great structure came to me from every imaginable source. Sometimes it came very unexpectedly and mysteriously."

"I certainly understand that," said Whitehead. "You see, when we are highly enthusiastic over something, the subconscious mind constantly leads us to what interests us. That is one of the great secrets of health, wealth, happiness and success. What is strongly desired, the subconscious realm of mind searches out and finds. I am under the impression that you became interested in the Great Pyramid and its prophecies because you are such an unusual person who was open to receiving information from 'other' sources. It had exactly what you were seeking even at your very young age—enlightenment regarding the present and future. It's quite natural that you were attracted by the mere mention of it. Isn't that right?"

"Yes, that does seem true," answered Michael, after a few moments of reflection. "All of my short life I have been much more interested in the present and the future than I have been in the past."

"That is as it should be," said Mr. Whitehead. " The past is dead and only the present and future are ours. To know something of Destiny's plan and program for the future means that anyone can definitely outwit tomorrow. Each must use his best reason and judgment about tomorrow or it will not only become his master but his executioner as well. Your father and I are of the old school, but we have both caught the vision, the inspiration and significance of the New Day to come. Through much thinking and a great deal of experience, which is by no means valueless, we are the equal of youth and with no great effort we can keep up with you and the rest of the younger generation on the journey "West."

Miles, delighted with the capable manner in which Michael discussed these things with Mr. Whitehead, had listened to the conversation without comment, but now interrupted, to say, "Do I understand you to say that age makes no difference once one knows the plans and purpose of the future?"

"Yes," replied Mr. Whitehead. "Age, environment, education and health make no difference whatever. Once a person starts turning

'West' you progress just as rapidly as you wish.

"In an out-of-the way place in my home state lives a charming old lady I have known intimately for some sixty years now. Somehow, we lost track of one another until one day a few months ago I accidentally ran across her. Of course, nothing worthwhile really happens accidentally. She was then over 85 years old and looked every day of it. Old and feeble, she was lifted out of her bed and seated in a wheel chair occasionally when she felt strong enough and was merely waiting for the end.

"Although her mind was quite hazy on most things, she remembered everything concerning the period of time when we were in contact and recalled scenes and even bits of conversation in which we had engaged and which I had long forgotten. She marveled that I had not grown old and decrepit and she just couldn't account for it. I explained that I had grown old but that I had 'come back', and that she, too, could 'come back' if she wanted to and I would be glad to show her how to do it.

"'No,'" she said, 'I have lived my three score years and ten, and fifteen years besides. Now I'm ready to pass on when the time comes.'

"'Why hasten the sunset?'" I asked her. 'You have sufficient income from your investments to keep you living comfortably. You can use the coming years to help instruct the masses during the dark days ahead. Although you have raised a fine family, you haven't contributed anything personally toward helping humanity. Why not start something spectacular in your own life? Your example will so amaze and inspire those who know you that they will be compelled to do wonderful things for themselves. You can become a shining example of what anyone can become during the period of time between now and even beyond the new millennium which is fastly approaching.'

"The idea immediately intrigued her and she began to see the possibilities of influencing others by her own efforts and actions. She realized that by improving herself in all ways, she would inspire others to follow her example. I promptly started instructing her on the five departments of life, the use of zeal, fervor, and enthusiasm as well exercising reason, will and judgement. I taught her the value of purpose and silence and a few other simple practices of a similar nature.

"This took place three years ago. She is now 88 years old and is no longer bed-ridden. Every day since our memorable talk she has made decided gains in every department of life. Eight hours rest each night, plus one or two in the afternoon, is all she requires. After a short time she abandoned her wheel chair for crutches; then the crutches for a cane; then she threw away the cane. Her mind has become keen and alert; she has developed a splendid sense of awareness; she talks before large groups at her church. Some time ago she travelled all alone to a large city a considerable distance away and delivered a lecture to a massive audience. You would have been so delighted and thrilled to hear her speaking. Her strength was amazing considering her former condition, and the audience took to the vital truths she expounded with so much enthusiasm it was unbelievable. She explained the five departments of life in a manner that even astounded me, and her discourse on zeal, fervor and enthusiasm, reason, will and judgment as well as the secret of silence was a masterpiece.

"I use this case to emphasize the fact that regardless of who you are or where you may be or what your condition is in any or all the five departments of life you can start where you are with and what you have and accomplish amazing things, providing you work with zeal, fervor, and enthusiasm and exercise reason, will and judgment."

"Have these things always been possible or are they just now becoming possible?" asked Miles.

"That question has a double answer, Mr. Miles," said Mr. Whitehead. "It is true that they have always been possible. But it's also true that they have been possible only for a few. It is now becoming possible for all human beings the world over but not all will accept them.

"After August 20th, 1953, there have been and will continue to be many signs and wonders. The Scripture speaks of 'signs and wonders which will follow those who believe.' Do you see that it states, 'follow those who believe?' This indicates that first one must make the decision to believe the teachings of the Great One and then the 'signs and wonders' will appear which will substantiate and confirm the beliefs as well as encourage the believer. Individuals will profit by them, but the masses will be misled. Only those who have decided to 'turn West' can hope to pass entirely though this period and enter into the Great Day when all good things will be the individual property of every person no matter what their station in life, from the least to the greatest."

Michael exclaimed in surprise, "Do you mean that there will be a difference in the standing of people in the New Dispensation? I had thought that this new era was to be one of complete equality among everyone."

"Many people are under that impression," answered Mr. Whitehead, "but it is not true. Never, in any dispensation or at any time in the future, can we be anything but what we are. Those who have already extended the five points of their star of life will be the greatest in the Golden Age to come. Those who have loitered and barely entered onto 'the Path' will be the least. What we build into our characters now goes along with us. And one who builds only a little will have only a little to take with him or her through the portals of the New Day. There is a minimum to what one must possess in order to enter the New Dispensation, but there is no maximum. That would penalize those who do the greatest work in assisting Destiny to usher in the new era."

"Yes, I can see that now," said Michael, meditatively. "I have been laboring under the delusion that equality among people was something to be desired, but I see my mistake. To give to the idle equally as much as to the industrious would be unjust. When I return to college I see that I'll have to correct some of the statements I have already made to my friends."

"I have one other piece of information for you to digest, Michael. *Never do for someone else what they are capable of doing or themselves.* I have already explained this to your father and now you must ponder this premise. You will gain additional insight and understanding of human relations once you realize the simple truth behind this most important statement."

Both Miles and his son meditated for a while on what Mr. Whitehead had just explained to them.

"Now I have a rather large question to ask," interjected Mr. Miles. "Would you give us some idea of what you expect to take place throughout the world after the 'Fifteen-Year Period'?"

"That is a very large question," smiled Mr. Whitehead, "and I can only touch on the highlights right now. In the first place, millions of people all over the Earth will be inspired with the larger vision of things to come. Each will be shown their own destiny. At first, one will think that whatever they may do in developing their five departments of life is something that they do of their own volition. This is not the case as they will discover later. Every person, even if only slightly interested in the matter of turning 'West' and helping others, has been directly or indirectly led into that endeavor by the 'White Forces' of good. That dear old lady I told you about thought she and I had just accidentally met again after such a long a time and that she had inspired herself to be young again and to do things for others. She knows now that this was not the case. For eighty-five years she had been prepared by the 'White Forces' to do a certain work. When she arrived at the place where she could set aside her own views and opinions, she was ready to be used directly by them. Immediately, an emissary, myself, was dispatched to her and I was very happy to be considered worthy of instructing her.

"Another case much nearer home is yourself, Mr. Miles. As long as you thought you 'knew all the answers' you were of no value to the White Forces. They were patient and aided you in your changes from year to year. Then one day *when the time was right*, I was sent to you. You had reached the place where you would welcome certain information that would benefit you and others.

"The 'White Forces' know each individual better that they know themselves. Unseen, they work with a person. They watch their aura. When it glows with a golden light the person is ready for higher instructions. The average person who is still one of the masses has an unpleasant appearing aura consisting of gray and somber and often with lurid red predominating. When one has started to hope for better things and begins to have a burning desire for them, the red changes to gold and they are ready to be taken in hand for instruction and faster development.

"Having already entered that age wherein such instruction is the order of the day, it is not necessary at first for a teacher to come directly to an individual. As soon as the proper amount of golden light is evident, instructions are given to them by various means. This usually begins to take place through the printed word. You think you've run across the information accidentally but these things are never accidental. They are purposely arranged and brought to the individual's attention. When the person is finally ready for an individual instructor, one appears. The teacher is not always in the flesh, as I will explain later. There is an old saying, 'When the student is ready, the teacher appears.'

"When the unsuspecting student has finished the course of reading, studying and absorbing the information, making it their own, and has started developing the five departments of life, a day will come when valuable individual instruction is given to them. This is done in a quite unique and interesting manner. At night, when the person is asleep, the instructions are impressed upon the mind. When one awakens in the morning exactly the information required is present. Often in the beginning, the impressions are not lasting and must be repeated many times all depending upon how sincere and open-hearted the student is. Eventually, the personal information which the 'White Forces' wish to impart to the individual comes through clearly. In the morning he or she wakes up with a perfect picture of the special and personal information which was conveyed during a night of instruction.

"After many days, they will wake up one morning and vaguely remember the place where the instructions were given. After still more time has elapsed they will remember the instructors themselves. So by this special method it is not necessary for the teacher to always appear in the flesh. Teachers themselves, who are in the flesh, are given instruction in this manner so that they can pass it along to those who cannot yet receive information during the night and bring it to their conscious minds when they awaken in the morning. You, Mr. Miles, and you, Michael, are of this type. You were both ready to progress but you needed someone to give you information directly in order that you would be able to quickly start your new work of helping others. Had you not been previously prepared anything I might have said would have been of little interest to you. It would have seemed fantastic or nonsensical, or even too good to be true. Also keep in mind that 'young men shall see visions; old men shall dream dreams.' What may seem a kind of 'vision' or 'dream' may actually be instruction given to the human being.

"Other individuals all over the world are now being prepared or are already receiving instructions concerning themselves and things to come. Some of these are being given information directly from certain individuals who are not of this world. They are instructed to pass on the information to those of their acquaintance who have shown genuine interest in advancing on the path of *expansion* and developing the five departments of life. They are located mostly in English-speaking countries. Every country has at least a few who have 'turned West' and many more people are being prepared for the same program. I hope what I have just said, gentlemen, clears up that point."

"Yes, it does," exclaimed the two Miles' almost simultaneously. Then after some comments by both father and son, Michael asked, "Will you give us some additional information regarding the date of November 27, 1939? The significance of it is not exactly clear to me."

"There are several significant dates in the 'one-hour' period," answered Mr. Whitehead. "Most important of all was November 27,1939. This is around the time that Hitler began in earnest his plan to take over the world by force and extermination. He was, in fact, very much involved in the black

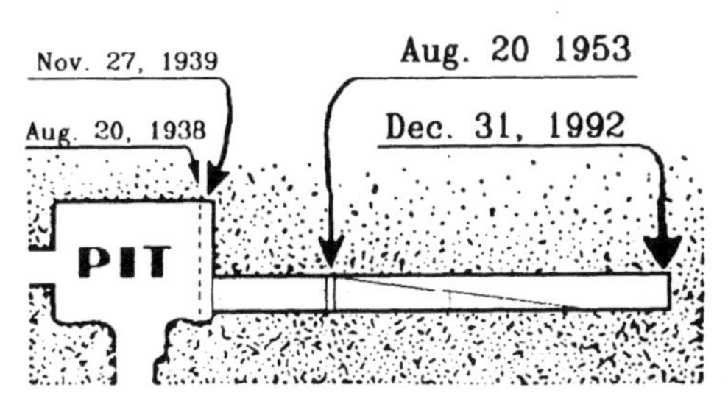

occult and demonology and felt he had been give 'power' by Satan himself. It was on this date that the Bottomless Pit was closed, so to speak, and the evil forces locked in and forced to move forward as time moved on and finally ending up in the Dead End Tunnel. It was then that these dark forces put up their greatest battle. It seems quite peculiar that, knowing they are fast approaching the end-times of annihilation, they continue to struggle against it. But let me explain further. Among the forces of evil are both personified evil beings, some of whom were cast down into the Earth when the great war in the heavens took place, and in addition many who have come into existence through man's own hatred. Anger, jealously, fear, remorse, envy and the like, all have created their individual 'demons,' so to speak, who spring from negative thoughts and energy. Each evil creature has created for him/herself a discoid, radiate form. It's no small wonder that they are able to fool many truth-seekers by appearing as 'signs and wonders' in the skies above nearly every country in the world, as well as in the heart and mind of the unsuspecting person. These man-made 'demons' plus the original evil spirits, beings and entities who are part of satan's angels are all doomed to be totally annihilated at the final ending of the world prior to the establishment of eternal peace for all who will live on Earth in love and harmony.

"But the day of their final extinction depends upon their united strength. Again, no one knows that day and time when 'the end' begins for these contemptible spirits who continue to hound mankind in every evil way possible in order to turn them away from universal truth. That is, it depends on how much evil force, through influence on human beings, that they are able to attract to themselves between now and the end of the ages. The more hatred, anger, jealousy, envy, revenge and similar negativity that they can generate in mankind before that day in time, the longer will they continue to exist. Like men condemned to die on the gallows, they seek reprieves and stays trying to delay the inevitable as long as possible. Remember their leader, the Accuser, (which in Hebrew means Satan who continues to accuse the disciples of Christ) still operates freely, going back and forth between

The Creator and mankind shouting unspeakable allegations against individuals who have chosen to walk 'the Path' and develop their five departments of life, living a virtuous and righteous life as an example to others. And in many ways through the work of sometimes unsuspecting individuals who think they are doing people a great service by espousing their own man-made teachings, causing people to turn away from the teachings of 'The Great Master', the more easily the evil ones are able to continue to add to their loathsome dominion. So many who might otherwise turn "West" are operating in a realm of delusion because of the teachings of so-called 'wise and inspired' men. How easy it is for the forces of evil to prey upon individuals by disguising themselves as the purveyors of something good!

"Good is immortal and evil is temporal; mortal. Evil is like the glowing heat in a piece of iron. When the iron cools off, nothing is left of the heat (evil). One single piece of iron will cool quickly. But if you put thousands of pieces of red hot metal in a pile they would remain hot for hours. If a million were heaped together, they would remain hot much longer. A huge mountain of red-hot pieces of metal would remain hot for years. And so it is with evil. This is why the dark forces personified in evil are putting up such a bitter fight. The more members they can add to their number from among the mass-minded of mankind, the longer will they continue to exist in the Dead End passageway. And so you can see that the future will be anything but dull, especially for the masses.

"The evil forces are quick to discover people who are part of the masses and who will do their bidding; small or great, it doesn't matter. If one can be used at all, the dark forces will find work for them to do. Today, wherever there are dictators, they are in league with the dark forces. Their lust for power when they were one of the common herd attracted the attention of the evil ones to them. Soon they became attuned to their influences and the day then arrived when they could be used. At first they did not suspect that they were nothing more than tools in the hands of a powerful dark force. They thought of themselves as clever and strong. When something fortunate happened while struggling for power over the masses, they considered it not only a lucky break, but one which was of their own making. Now these dictators are just puppets of a powerful evil force and if they tried to resist the will of evil they would instantly be destroyed by the those who put them in power in the first place. They are now an integral part of it. The evil autocrats are the ones who should fear that the dark forces will not need their services any longer. And they have good reason to fear. The moment evil finds other human beings more suited to their purpose, present day dictators will be flung to the side and replaced by even more evil men who will do anything to keep their positions of power intact.

" In America, we are fortunate not to have accepted a dictator. At least not yet. But the danger still exists. We must work hard and ardently to prevent such a calamity from ever taking place.

"Did you know that the President of the United States has the power to issue executive orders that would place every person under an absolute dictatorship? These orders allow the take-over of just about every phase of life as it exists today, from the media to health; welfare and education; airports and aircraft including all other means of transportation; housing; and requiring a work force to be formed and operated under the direct authority of the government as well as requiring all persons to register with the Postmaster General. Already, all male students are required to register for the 'draft' (of which none exists today) when they become eighteen years old. In addition, FEMA (Federal Emergency Management Agency) would be given authority to oversee that all Executive Orders take effect upon issuance by the President during times of increased international tension or economic/financial crisis. There are eleven Executive Orders in all. It's a frightening thought how quickly the United States of America, the 'home of the free' and the 'land of the brave' could so easily and quickly become just another dictatorship in the world.

"I have another piece of information which may be of interest to you and to all who strive to learn the truth of the secrets of the universe. On October 5, 1982, Dr. Brian T. Clifford of the Pentagon announced at a press conference ('The Star', New York, October 5, 1982) that contact between U.S. citizens and extraterrestrials or their vehicles is strictly illegal.

"According to a law already on the books, Title 14, Section 1211 of the Code of Federal Regulations, (adopted on July 16, 1969, before the Apollo moon shots), anyone guilty of such contact automatically becomes a wanted criminal, may be jailed for one year and fined $5,000.

"A NASA administrator is empowered to determine, with or without a hearing, that a person or object has been "extraterrestrially exposed" and impose an indeterminate quarantine under armed guard, which could not be broken even by court order.

"There is no limit placed on the number of individuals who could thus be arbitrarily quarantined.

"The definition of 'extra-terrestrial exposure' is left entirely up to the NASA administrator, who is thus endowed with total dictatorial powers to be exercised at his slightest impulse, which is completely contrary to the United States Constitution. The legislation was buried in the 1211th subsection of the 14th section of a batch of regulations very few members of congress probably bothered to read in their entirety and was slipped onto the books without public debate. With this action, the government of the U.S. has created a whole new criminal class: unidentified flying object/extraterrestrial contactees. It appears that the primary effect of this law is not to prevent contact, but to silence witnesses.

"According to a NASA spokesman, Fletcher Reel, the law is not immediately applicable but in case of need could quickly be made so. That means that the law is subject to interpretation and could be effected at the whim of a NASA or government official. All this was done by a government which still to this day refuses to even admit the existence of unidentified flying objects or the truth of visitors (teachers) not originating from planet Earth. Both of these are 'signs and wonders' which *follow* the believer. When one uses his Spiritual mind in combination with the Mind of God, one will see and discern the truth of these 'signs and wonders'. In addition, the malevolent ones use these same means to confound the masses and create confusion whenever they can. Many of the airborne objects are of wicked doing to fool the unbeliever. Again, when it comes to 'signs and wonders' test the spirits. Do not be caught in one of the poisonous traps set to turn you away from the Spirit of truth.

"This is very interesting, indeed, and knowledge which nonetheless you may find helpful at some point in your future endeavors of bringing to light the universal teachings which have already been imparted to you and the scholarly principles which will later be delivered to you.

"I give you this information so that you will be aware that the White Forces consisted originally of brothers and sisters (angels) from far away lands who were sent here under direct orders of the Great One to help the human race develop into what Destiny had intended for each of them. They in turn have trained certain individuals including myself to assist in this work. Remember when I told you about a special place in the Andes where I was invited to visit and learn? Who do you think those 'teachers' were? There are many secrets of the universe which are now being brought forth by those who have been given the courage in the face of personal adversity to bring forth the teachings and messages from these special ambassadors of truth.

"Groups have been formed around the world so that the information could be circulated in an orderly fashion. Each group is called the '*Inner Circle*'. Many have accepted these lessons in full, others have accepted certain portions of them and still others are questioning whether they should continue to be involved in such a group at all. Each must decide for him/herself. Just as each of you have had to make your own decisions as to if and how you would continue on the Path, it is a personal decision for each individual which path to follow and one which should not be made hastily but after much thought and questioning of one's self and their future.

"Now, back to our original subject from which I digressed. As we advance in time, the strife among the masses will increase in intensity. This is the work of the damnable dark forces. But not everyone will be involved in this struggle. Those millions who have chosen the 'Westward Path' are safe from the dangers of what will ultimately be the downfall of the mass-minded ones. When the five departments of life are well developed each individual will instinctively know right from wrong. Those who are representatives of the White Forces will be well taken care of. When you know what to expect and how to cope with each emergency as it arises, there is nothing to fear. The ones who have 'turned West' are working with the White Forces and will definitely know how to Outwit Tomorrow today. Remember that the purpose of the 'one-hour' period is to rid the world of as much evil influence as possible and make it a fit place for all of humanity where each person will constantly improve in their efforts to become a greater and still greater human being.

"There is no necessity for me to give you anything further regarding the future. For those who will not turn 'West' the future is so dark that I won't burden your pure thinking with that information. For those who are *Outwitting Tomorrow* by their deeds, thoughts, and actions, the future is so very bright. Be careful not to allow yourself to start imagining and wishing to such an extent that your present work is entirely ignored. You must continue to study and learn as much as you can if you are to make substantial progress."

Mr. Whitehead stopped, waiting for Miles or Michael to speak.

After a while Miles spoke up, "I have a personal question," stated Miles. "What do you suggest that I do during the coming months in order to be better prepared for working in unity with the White Forces?"

"I suggest, Mr. Miles, that you stay right here at home until after Michael has completed his studies at the University. During that time we will keep in touch with one another. Make every effort to improve your five departments and fill in the points of your star of life. Be sure to pay close attention to the financial department. There are several opportunities you are overlooking which could generate a tidy income for you and your wife. Start mixing with people so you can find those who are ready to turn 'West". Try to realize that you are working with and not for the White Forces and they are actually working through you. Know that if you do little things well, greater things will be given you. With these greater things will come greater joys and thrills and also enormous responsibilities.

"Live constantly in a state of expectancy, so that when the unexpected occurs you will not be surprised. Lack this quality, and it is hard for the White Forces to contact you. It is hard for some people to see coming events and therefore the resulting changes which are taking place in their lives.

"What I have just said to your father is also for you, Michael. Live in a constant state of zeal, fervor and enthusiasm. Always be cognizant of what is going on around you. In addition, it is quite necessary for you to return to college this fall. There is a great amount of work that only someone with your personality can do. I want you to disseminate as much information among your classmates regarding the things to come as you possibly can. Do so with zeal, fervor and enthusiasm while at the same time exercising reason, will and judgment as you discern carefully those who will be receptive to what you have to say."

"I most certainly will," replied Michael. "I have been doing some work with them already, but I have had to proceed cautiously so as not to be accused of preaching at them. Now that my knowledge and understanding have been increased I'm sure I'll be able to do something really worthwhile among my fellow students. It is amazing how many have become interested in this subject during the past year, including people whom I never would have thought of as having a serious bone in their bodies."

"Living, as we are," said Whitehead, "in the time period when people either become interested or they do not, it's not at all surprising that we find people everywhere, in all classes, creeds, ages and environments whose minds are open to receiving the teachings to which you have been privy. The vibrations of the new Dispensation make this possible. The moment you graduate, Michael, there is a tremendous task awaiting you. And so, gentlemen, that is all for now."

"That seems to be quite sufficient to keep us both busy for a considerable length of time," commented Miles.

"It will be an invaluable and an intensely interesting period for those of us who work zealously," said Michael.

"You both are right. It will be both interesting and busy," agreed Mr. Whitehead. "And now, unless we hurry, we're going to miss the sunset on the ocean."

All three got up and before long they were climbing the trail to the summit. They sat down to rest and waited for the sunset to arrive. Mr. Whitehead utilized this time to run over the various points on which he had instructed Miles and Michael.

"1. There are five departments of life, symbolized by the five-pointed star. They are the Spiritual, Mental, Physical, Social and Financial.

"2. The powers that cause us to do great things from little beginnings are zeal, fervor and enthusiasm.

"3. The Secret of Silence enables us to retain enthusiasm for our aims and plans.

"4. In order to dissipate unwanted desires, we talk them out or use the writing-out exercise.

"5. All the good we are doing and we are going to do must be done in the now, exercising reason, will and judgment in everything.

"6. Knowing that both great and terrible things are coming, we should not be fearful of the future if we remember that we are working with Destiny.

"7. We must live in a constant state of expectancy, anticipating many good things. We will live in this frame of mind so absolutely that we will not be surprised when good things do come, but will be exceedingly thankful.

"8. Realize that everything we do and think is inspired by the White Forces.

"9. We will do what we can, wherever we are, with whatever we have.

"10. Expect great changes in every department of our lives.

"11. We will work secretly as well as openly, realizing that a good deed done secretly often accomplishes much more that a larger one done publicly.

"12. We welcome information, understanding and teaching from every source, knowing that whatever is good is brought to us by the White Forces and we will use our gift of discernment to determine the value and truth of all sources of information.

"13. Eliminate from our minds fear, hatred, revenge, jealousy, bigotry and all such evils. They cannot be carried into the New Day.

"14. We will always remember that love is the greatest force in the universe for good in the world. Although we are incapable of loving everyone now, we will gain in ability as we push out the five points of our individual star.

"15. Always remember to work with the 'construction gang' and never with the 'wrecking crew.'

"16. Faith and a good conscience are your weapons against evil. Because you have been tempted to turn away from The Path and have chosen to stay on course, you are able to help others who are tempted do so.

"17. Encourage one another. Do not allow your heart to become hardened by the reactions of others whom you are trying to teach these lessons for their own benefit. How weak are those people who refuse to listen, contemplate and grasp these lessons!

"To quote Anne Frank, 'How wonderful it is that no one need wait a single moment before starting to improve the world', and you too can do this one person at a time."

He broke off abruptly to say, "Look, the sun is about to sink beyond the horizon."

Michael knew what to look for. His father had described the afterglow to him. As the sun disappeared below the horizon, he gazed intently at the spot where it had been.

"There!" he exclaimed, as he saw the flare up. "The sky is brilliant again."

Then Whitehead spoke softly, "Each of us must go his separate way now. There is an old German Proverb which states, 'If someone goes on a journey, he's got something to tell'. But a year from now, if each of us has done his part in helping humanity and has been consistent in his own self-improvement, we shall meet again to receive further instructions from the White Forces concerning our activities in the future."

Filled with thoughts of the parting and of the role each was going to play in *Outwitting Tomorrow*, the three men proceeded silently down the trail toward the highway.

CHAPTER 12

A Shining, New Life

A year had passed since the three friends parted, now the time came when they would meet once more and unite in their mission of helping humanity.

For Mr. Whitehead, it had been an intensely busy year. He had devoted much of his time to teaching and an equal amount to learning. The path from "a clod to a god" is a long one and the saying, "onward and upward forever" aptly describes it.

Whitehead had spent a part of the year at The retreat in the Andes in South America. He was far enough on *the Upward Path* to enjoy the invigorating quietness of a few short months at the great institution and he benefited significantly from his stay there.

He was looking forward to the time when men like Miles and his son Michael would be sufficiently *expanded* in the five departments of life, Spiritual, Mental, Physical, Social and Financial, to receive an invitation to visit there and receive instruction from them and even teach a class in some subject in which they were especially versed. Even a Master must take special courses in certain subject matter in which an ordinary person may be highly versed.

As for Miles, he had made great strides in developing every department of his life. No longer was he the drab, sour, stooped old man of the previous year. Instead, he had become a most magnetic personality. His manner and attire radiated confidence and ability.

After exhausting every reasonable means of helping his friends and neighbors, he was forced to acknowledge the fact that while they were interested in the personal changes he had made, they also were envious of them. They resented the fact that they had not been selected to experience the good fortune which had come to him. As a consequence, they rejected his teachings. But the seed had been planted and the soil was not entirely barren. Some day the seed would spring into life and grow. It was only a matter of the right time presenting itself.

In contrast to the attitude of his friends and neighbors, Miles had become quite a personality with the guests at the hotel in Casa Del Rey. He was urged to give talks and did just that with ever-increasing success. Truly, "a prophet is not without honor save in his own land."

Miles had enjoyed a very prosperous year after discovering several new ways of earning more money. Mrs. Miles and the younger children all wanted to move into Casa Del Rey now that they could afford it. But Miles discouraged the idea saying, "Let's wait until Michael comes home and hear what he thinks about it."

At the University, Michael had been actively involved. His broader viewpoint on life had brought about great changes in him. Although the last year of college was considered a particularly difficult one, he managed it easily. In his spare time he instructed a group of students who were interested in the new day dispensation. They called their class the "C.S.T.", which meant "The Circle, Star and Triangle Philosophy." He was amazed to find that those who had not been too serious about life were much more open-minded than those who thought of life as a serious, direful, and awe-inspiring existence. And those people who didn't seem to have a single serious thought in their heads would turn out to be the easiest to interest and would obtain the best results. Those who were sure of themselves and knew "all the answers" had not even heard of the "C.S.T. Club."

The Miles' gave a big homecoming party for Michael and Mr. Whitehead, who both arrived at Casa Del Rey the same day. It was such a joyous occasion for all concerned. They were all so fond of one another and were utterly delighted to be together once more.

The next ten days were happy and carefree. The three men, Whitehead, Miles and Michael, drove around the vicinity in Whitehead's car. Michael did most of the driving. Never before had he experienced the thrill of so much power, pickup and speed in an automobile. They also did a lot of mountain climbing. Miles enjoyed this to a greater extent than did the other two. This was mostly because not even once did he feel the slightest twinge of pain from his old-time rheumatism or other joint aches and pains. They spent considerable time sailing on the blue-green waters of the Pacific. Their conversation was spontaneous and fascinating.

Mr. Whitehead had seen so many people make just such a change in themselves as had Miles, but he was amazed at how quickly Miles had accomplished the change. He also was also quite pleased with the progress that Michael had made. Physically, Michael was much the same, but he had taken on a much more carefree manner and was far less sensitive.

Whitehead commented on it one day, saying, "The moment I saw you, Michael, I knew you had made a lot of progress in a short period of time and like your father, you had taken great upward strides. Maybe you are wondering how I could have known this. Well, to tell the truth, *I saw it.* I am gifted with the ability to see and read auras and a year ago yours, while even then very fine, could in no way compare with the highly-colored one you now radiate. It indicates a well-developed personality in all five departments of life. Tell this to no one and beware of false pride. You don't want your 'gold' to turn to 'brass'."

Later that same evening all three gathered in Mr. Whitehead's suite at the hotel.

Settled comfortably, Whitehead said, "We have enjoyed a few days of complete relaxation. Now we must prepare for a period of intensified action. There is a lot of work ahead for all three of us. I have been instructed by the White Forces to inform both of you that you have an opportunity at this time to do a greater work for them and with them. If you don't care to take advantage of this opening there will be no penalty for your refusal other than it will be an opportunity that will be lost, never to be regained. I tell you this so that whatever your decision is regarding the proposition I am about to make to you, it will have been made of your own free will without any fear of consequences. Are you ready to hear it?"

"We are," chorused Michael and Miles.

"Very well," said Whitehead. "Next Saturday evening I am leaving by plane for the northern part of the state and from there I will be flying to Washington, D.C. Something is happening there that might prove exceedingly disastrous unless we act now. Unless something is done about it soon, the black forces in the 'Dead End' passageway will be permitted to wreak havoc among our congressmen. If this should happen, chaos will fill the land. Unless we come to the aid of the White Forces, who even now are stemming the tide against evil principalities and powers, the damage will have been done and there will be nothing we can do to change it.

"I will be gone at least a year, Mr. Miles, and I want you to go with me next Saturday and help in this work, but only if you feel called upon to do so. You will have until tomorrow at this time to think it over and decide. Discuss your thinking on this subject with no one, not even with your wife or your son, Michael. I suggest that you spend tomorrow alone at the ranch while you make this momentous decision."

Mr. Whitehead's proposal left Miles speechless with amazement. Then many questions and worry filled him. How could he leave on such short notice? Who would take care of the family, the ranch, the poultry? The very thought of traveling by air caused beads of perspiration to break out on his forehead. But with concerted effort, he overcame his emotions and answered, "I'll definitely let you know my decision tomorrow evening."

Whitehead nodded assent. Then turning to Michael, he said,

"I have two propositions for you. First, if your father decides to go with me you are to take his place at home and here at the hotel instructing the guests who are ready for the teachings which the two of you have already received. The second proposition is open to you only if your father decides not to go with me."

All the next day at the ranch, Miles was in a mental quandary. Should he go with Whitehead or not? He considered it from every angle. One moment he felt that he should go, the next he decided against it. Realizing that he was not analyzing the problem in the way of a master, he composed himself and sat down in the big easy arm chair at the "hut" and relaxed. Soon he fell fast asleep. Gradually, as he began to regain consciousness he sensed that he was passing through space, high above the Earth. In time, he reached a point which looked directly over the Capital Building in Washington, D.C. The entire place appeared to be in an inferno. Lurid, red flames shot up at intervals through the murky turmoil. For miles around the dark forces were gathering. Even in the ground beneath the Capital there was a seething mass of diabolical activity. It was a terrible and ghastly sight, and the thought of what was occurring was even worse to consider. If only the White Forces could do something, he thought, the impending danger could somehow be lessened. At that moment he heard a *voice* within him say, "Look up!" He did so, and to his astonishment, the whole heaven above was brilliant with the light of the *White Forces* gathered together in one place.

"Why don't they do something," he thought.

Again the *voice* within spoke.

"They only help mortals by working through individuals."

He was wide awake now and recalled every detail of that scene remembering what the *inner voice* had told him. When he saw Mr. Whitehead that evening he announced without hesitation,

"Mr. Whitehead, I am ready to go with you. I will have my affairs in order and I'll be ready to leave on the evening plane Saturday just as you suggested."

Whitehead was delighted with the decision. "I had a strong feeling that you would accompany me, Mr. Miles," he said. And then turning to Michael asked, " What have you decided to do, Michael?"

"I have decided to stay at home and take Dad's place, both here and at the ranch."

"Splendid," applauded Mr. Whitehead. "But what influenced you to make that decision?"

Michael responded, "I seemed able to look into the future and I saw you and Dad going east and myself taking over Dad's work. Then I saw people in great distress coming to me here at Casa Del Rey for relief and I had something to give to each of them to alleviate their suffering. The moment they accepted it their anguish disappeared and they went away joyous and happy."

"Old men shall dream dreams and young men shall see visions'," again observed Whitehead to himself, softly. "Truly, we are at the gate of the New Day."

Then gazing at both Miles and Michael intently for a moment, he said, "I have the great pleasure to inform you both that you have successfully passed the test which will permit you to do a still greater work for humanity. Had you not worked diligently and expanded evenly in every department of your lives during the past year, you would not have made the decisions that you have. We must always remember that the work we do takes an equal blending of heart and head in order to carry out our missions to enlighten, inform and instruct. We must keep the heart flame uppermost in our thinking because the flame represents the mind which is at the base of all our activities.

"At this time, I wish to also introduce you to a special ceremonial prayer of protection which was given by certain ones of whom I have previously spoken and for the express purpose of providing you with a Spiritual place of refuge and grace. This should be passed along to all who will listen and accept the underlying truth behind this solemn prayer. It is called *the Ring of Fire* ceremony prayer.

> *'Eternal Father, creator of the universe,*
> *hear this day my petition.*
> *Surround me now with your divine "Ring of Fire,"*
> *the fire of your protection,*
> *the fire of your abundance,*
> *the fire of complete healing,*
> *the fire of divine abundance.*
> *I command the hand of almighty God on my behalf.*
> *Let it be so, this very moment,*
> *in the blessed name of*

our Lord and Master, Jesus Christ. Amen.
So be it!'

"This prayer is for the specific purpose of providing protection for you from outside forces which may wish to obstruct your vision and actions on behalf of the forces of good. It is called the 'outside ring' of the *Ring of Fire*.

"You are instructed and advised to pronounce this prayer each and every morning before you begin your 'work' and each evening before you go to bed. Many others also recite these words before leaving for a trip, say in a car, an airplane or the like. There are records of people's lives actually being saved and serious, life-threatening injuries prevented through the use of this information. Later, I will give you further directives concerning other, similar prayers. But for now, begin to use and teach this one." (See Appendix A, Expansion)

It was late afternoon and Michael, having bid his father and Mr. Whitehead a hearty good bye earlier in the day, felt strongly compelled to watch the sunset from the old familiar spot on top of the ridge. He reached the summit and was viewing the path of golden light cast across the ocean by the low-setting sun, when he heard in the distance the drone of powerful airplane motors. Louder and louder they sounded and from the south appeared a north-bound aircraft, flying lower than he had ever before remembered. It was so close that the wings seemed almost to touch the mountain. Michael's quick eye caught a glimpse of his father and then the familiar face of Mr. Whitehead as they passed. They waved as the plane dipped its wings from side to side several times in the gesture known to pilots as hello and goodbye and gradually the plane became just a speck in the sky. In the distance the sun dipped below the horizon . Then came the "flare-up."

Watching the afterglow, Michael realized that by traveling westward fast enough one might overtake the sunset and there would never again need to be darkness. The thought fascinated him as he began to realize that for all those who "turn West", *there is no more night.*

✪

Appendix A

The 'Inside' Ring of Fire Prayer

As previously promised, the following is the *Ring of Fire* ceremony prayer for the "inside" of the ring. Remember, that the ring has two sides, inside and outside. The inside ring of fire is a prayer for you as an individual to assist you in your walk on the Path from a 'clod to a god'.

Eternal Father, creator of the universe,
hear this hour, my earnest prayer.
Anoint my lips, my ears, my mind and heart,
with your divine flame of fire.
Provide me now with
perfect health, longevity, and abundance.
Revitalize and redirect my energies into divine service.
Touch me with your divine inner flame of fire and
cause me to be the master that you intend me to be.
In the blessed name of the master and creator of the universe,
the Lord Jesus Christ
so be it.
Amen!

The same admonishment applies to this prayer as to the Outer ring prayer. Recite it daily for your own benefit and watch the changes happen for you.

Mysteries of Creation

The entire Trinity is represented in creation. The *Father* is the first cause of all that is. God was before everything existed. God exists eternally and infinitely before anything else was. He is a personal being. The *Son*, the powerful *Word*, is a moral being through whom God created all things, all things created by him and for him. After each "day" of creation, God saw/said it was "good", meaning 'without sin'. The word "created" in Hebrew means "activity only God can do". And "day", in Hebrew 'yom', is loosely interpreted as 'eternity'. When the Bible refers to the days of creation, a length of time, a period of time, it doesn't really matter the exact period of time. The point is that God, the Trinity, created all. When the Son created, it was matter and substance that didn't exist before. It was without form and void, empty, baron.

The *Spirit* of God spoke the world(s) into existence. In Genesis 1:1-6, Jerusalem Bible, "In the beginning (in Hebrew 'out of chaos') God created the heavens and the earth. Now the earth was a formless void, there was darkness over the deep, and God's spirit (The Holy Spirit) hovered over the water.

"God said, 'Let there be light,' and there was light. God saw that light was good, and God divided light from darkness. God called light 'day', and darkness he called 'night'. Evening came and morning came: the first day.

"God said, Let there be a vault (from the ancient Semites the 'arch' of the sky was a solid dome holding the upper waters in check) in the waters to divide the waters in two."

There was in fact an ice canopy which surrounded the earth and thereby maintained a constant temperature on the surface.

God brought an orderly universe into existence out of primordial chaos. And the sounds of the heavenlies could be heard singing the praises of the Creator.

In chapter one of Genesis, verse 26, it says, "Let us make man in our image, after our likeness...God created man in his image, in the divine image he created him, male and female he created them..." This is a mystery because in Genesis, Chapter 2, there is another story of the creation of Adam and the subsequent creation of Eve. For further clarification on this mystery, you may refer to my audio tape entitled, "The Mystery Land of Quello". Also, see Jeremiah, Chapter 4 for additional information.

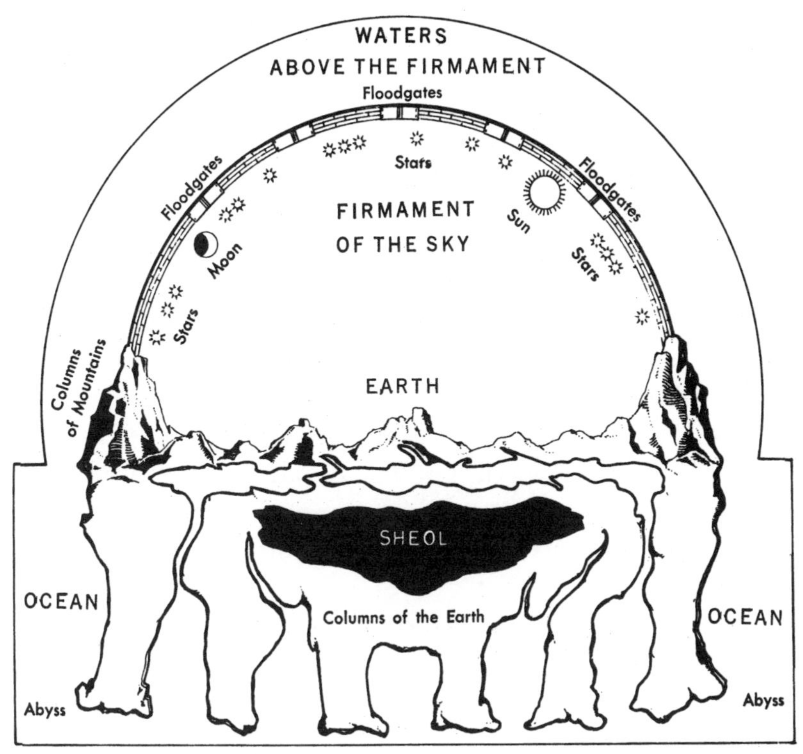

THE WORLD OF THE HEBREWS — Graphic representation of the Hebrew conception of the world. God's heavenly seat rests above the superior waters. Below these waters lies the firmament or sky which resembles an overturned bowl and is supported by columns. Through the openings (floodgates) in its vault the superior waters fall down upon the earth in the form of rain or snow. The earth is a platform resting on columns and surrounded by waters, the seas. Underneath the columns lie the inferior waters. In the depths of the earth is Sheol, the home of the dead (also called the nether world). This was the same prescientific concept of the universe as that held by the Hebrews' pagan neighbors. (Reproduced with permission from THE NEW AMERICAN BIBLE, SAINT JOSEPH EDITION, Copyright © 1992-1970 Catholic Book Publishing Co. New York, NY. All Rights Reserved).

Because Adam was created in the image and likeness of God Himself and Eve likewise, it stands to reason that there are both male and female aspects to God.

When God created the beautiful Garden of Eden, there were two special trees within the garden. One was the Tree of Life from which Adam and Eve could eat. The other was the Tree of Knowledge of Good and Evil from which they were forbidden to eat. They were told that they would surely die if they did so. This meant that they would become subject to the process of aging unlike other original creations (angels) who *never* age or die.

There soon came a serpent, Lucifer, the son of the morning, the most beautiful beast of the field. He stood upright and began to use his wiles on Eve telling her that she wouldn't die if she ate from the forbidden tree, but would be just like God. Nothing happened to her at all when she did eat. The change came when Adam ate the forbidden fruit. For it was to Adam that God had given the admonition not to eat from that particular tree. When Adam ate, the serpent became a snake slithering on the ground and both Adam and Eve fell from the Grace of God. This happened because of Adam's disobedience. Now they saw themselves as naked, their eyes were opened to feelings and emotions that they never knew before and they felt guilty for being unclothed.

Because of their guilt, they tried to hide from God but could not, of course. If they had been able to get back to the Tree of Life, they could have been restored to their previous state. But instead, they were cast out of the garden and you know the rest of the story.

Divine/Human Relationships

I know that this subject will not be "too heavy" for most of you so I will continue by sharing these thoughts with you.

According to Earth laws, man enters into a relationship with another individual. Because of the principles that all of us have already learned from The Master Creator and which have been told to you here in this book, we are to follow all of His teachings. You must seek to apply them to yourselves on the human level. Then learn how to use the Divine principles in your own approach to living life on a human level in order to seek to elevate yourselves to a much higher plane of Spiritual understanding in your relationships with others.

This Divine principle of The Creator's personal care and relationship with each individual, when applied to your own lives, allows all men and women to also enter into a relationship with each other according to Spiritual (Divine) law. You might also say that the Divine principles enter into each man's and woman's heart. Therefore home life, as it is known, should reflect a form of godliness without which the residence can represent a jungle of confusion.

Upon careful examination, please note that life is an interplay of the positive and the negative forces which exist. Matter and energy enter into a relationship with each other under "provocation" which is identified as stress and pressure.

A man and his wife enter into a relationship with the same challenge and excitement. After the destruction of the land known as "Quello" and during the beginning of The Creator's second attempt of populating Earth with man, Adam "knew" his wife, Eve, and she bore him a son. The word "knew" actually means, "to enter into a relationship with."

Likewise, when one enters into a relationship with The Creator, there is a mutual respect and understanding that overcomes every obstacle. The evolved lifeform always serves the purpose of "Divine life."

When you hold resentment, fail to be patient or neglect to follow the obedience of love, you are slaves in bondage to creation and to the power of "provocation." When you hate, you will soon discover that you are unwittingly clinging to negative powers and what is called "disobedient excitement."

In reality, man and woman are expressions of the Oneness of God. Woman seeks unity with man and man should seek unity with woman. This is the manner in which this plan originated at the beginning of life on Earth.

Animals do what they wish to do, any time they wish to do so, without any fear of punishment. Man, on the other hand, needs to take a few steps backward for the purpose of examining his actions and motives prior to making any further moves.

In this 21th century, this new millennium, there is too much going on between adults that should not be. There are both men and women who covet the mates of others. There are no rules set is the animal kingdom except for those that they create for themselves. However, as human beings, there is going to be a high price that will be exacted from those who are guilty of possessing "roaming eyes and ideas."

When you recognize that a man or a woman is already committed to another, either by word or by marriage, others should beware lest they find themselves breaking Universal laws by their own thoughts as well as actions.

Those from the outer reaches of space have seen this occur on the planet Earth over and over again. There is absolutely no excuse for you, as an enlightened being, to violate these Laws of God for every self-respecting man or woman by a lack of "self control." Those in the so-called "new age" movement have shown an utter disregard for the sanctity of the lives of others under the cloak of "enlightenment" and this has created a situation in the lives of many that demands immediate attention. Be cautious with whom you associate. Use the gifts which have been given to you to discern the true Spirit of so-called "enlightened" ones. Treat others as you wish to be treated. Be careful as you might be entertaining angels unaware.

Please continue to study the various works which have been presented to you by certain chosen ones who have been trained in all manner of subjects. The relationship between individuals and their Creator is one of utmost importance and must be cultivated on a daily basis if it is to grow and prosper. It is the "Great True Spirit" which draws each of you to Him and He will not only teach you but also bring to remembrance all that you have read and learned before so that you may apply it to your own life and also pass it on to someone who is not as far on the Path as you might be.

I know that this perhaps may not be the type of message that you would expect to read here, but those who have come here from afar and occupy the ship, Victor One, which is stationed outside Henderson, Nevada, in the United States, feel that the time is right now for me to speak out regarding various human subjects and that you, the reader, are ready to listen and absorb these universal truths.

Hate

Hate *must* die daily. It must be completely eliminated even from your vocabulary, let alone from your heart. If it does, *love* will spring forth and along with it patience which will live long and flourish. Your daily practice of the full exercise of the true Divine Love along with the pronouncement of the "Ring of Fire" prayers will ensure that you are on the correct pathway. If you follow this pathway, meditate and seek His Kingdom *first*, you will find a place of existence in which you may enjoy the pleasures of oneness with the Creator. Hate is a destroyer of anything and everything good, while *love* is the fire whereby eternal bonds are established. You *must*, at a cost which you do not wish to pay, eliminate any hateful thoughts and deeds from your life. Those who hate are operating under the total influences of the dark forces and you would do well not to associate with those individuals in any way. Hatred can so cunningly creep into your life when you suffer some kind of rejection, physical or mental assault or something similar. While you don't have to like what may have happened to you or to one you love, take special care not to let hate fill your mind and body with its poisonous tentacles. It can be so powerful and overwhelming that your very soul could be in jeopardy. Replace hateful thoughts with those of a positive nature toward yourself and for your betterment. In this instance, disregard the individual who has wronged you and remove this revolting emotion from *your own* heart. Forgive *yourself* for having these thoughts and move forward with love of self without any condemnation for the hateful notions which you may have had.

Divine Relationship

This is a relationship for which you must all strive. It is the relationship that will cause you to be "holy" and acceptable in His Sight. When you have conquered your "worldly" desires, you will be able to strive toward that mark that has been established by the King of Kings. That mark is certainly high and above any that you could possibly attain on your own. Therefore, there must be a shedding process which includes doing away with the "old" you and replacing it with the "new" you so that you can qualify to run this race of Faith. Shed animosity, pride, lust, jealousy, etc., and begin to run that beautiful course toward *expansion* and everything that it represents and which will, in the end, give you the prize that is in the highest calling of The Master.

Resentment will continue to rob you by preventing your free exercise of virtue and loss of confidence. Sometimes you unconsciously resent those whom you need the most. The Creator has been "blamed" for more things than I care to mention here. Resentment and blame only create more need and to be needy is not an expression of true love. Others will see this and call it weakness and resent you for that thereby placing in jeopardy the position you have attained as a leader and teacher. Only *love* and *patience* can detach you from the "natural law." Hurting others in response to your being hurt by someone else creates an unending circle of negativity. Respond in any and all circumstances with *pure love* that comes only from *Him*. Keep quiet when the occasion dictates. He and He alone will see you through whatever adversity may afflict you. You must work hard to establish and nourish your own personal relationship with The Divine. Through this alliance you will be able to operate in a spirit of true love and care for your fellow man.

Attracting Helpers

Everyone has friends. You know them as guardian angels, ministering angels, guardians, etc., and they are a gift from above. Some children are even taught to give them a name to more easily personalize them for the individual. And those more advanced on the Path may intuitively know the name or names of their angels. They are assigned to watch over and to benefit you individually. And remember that they are not here to do everything for you. You must do it for yourself. They will not answer questions when you can get the answer for yourself. It bears repeating here that you should *never do for others what they are capable of doing for themselves.* This is a lesson which should be at the forefront of your mind in all situations where you are required to make a decision regarding another individual. You would do good to learn and incorporate it into your daily living. Expand all five departments of life evenly and seek to be in balance at all times. To depend on anyone else to do it for you is not good. Exercise your free will and make your choices carefully after meditating and praying for answers. They will be given to you in one of the three ways which The Creator speaks to His people—visions, dreams and revelations. Your angel helpers are part of these ways coming to you in your dreams, sometimes giving you revelations regarding a subject or situation with which you need assistance and even being a part of some type of vision which you may experience. Sometimes, a vision is something you see with your physical eyes and sometimes you will see it with your mind's eye. It could even be a fleeting image in your mind. Know that your own personal angels are constantly at work on your behalf and in your lives.

The Truth About 'Soulmates'

For many hundreds of years, the dark forces have set up all kinds of "organizations" with whatever kind of belief and doctrine that would please the masses. This was done because one of the techniques that the dark forces use is half-truths and lies to the mass-minded thereby inoculating them against the whole truth with lying wonders and signs. The masses fall for it in great numbers. This "lines the pockets" of the stooges who sell these lines at "so much per."

The doctrine of soul mates is found nowhere in the Bible. This doctrine supposes that everyone has someone somewhere who is their "other half", their "soul mate" (opposite sex of course). There are many married persons who are not very happy. This gives an incentive for the unhappy person to look elsewhere for their "other half", thinking they are with the wrong individual. If these poor people would expand their tree of life (as previously referred to earlier in this book) instead of searching for their "other half", they would be less dissatisfied with their present state of affairs.

It is most difficult for "professional searchers" to follow the golden cord which keeps one attached to the Love of God if they are bouncing from one mate to the other every time the whim occurs to them. These "soul mate sleuths" often have to have "additional money" to continue the search. Yes indeed, this has become quite a business venture for many so-called psychics and seers. Beware of those who are out to separate you from your hard-earned money by promising to find that special "soul mate" which you are seeking. They will tell you anything you want to hear and support whatever belief or position which you have chosen to take at any particular time in order to "reel you in."

There are those who have received the gift of "sight" from the One True Spirit. Remember to test the "spirits" to know their source.

"Soul mates" as talked about on Earth is a farce. There is no such thing because people place a different connotation on the word than what it means. While you are busy searching for your "other part", you are not expanding in your five departments. If you are lonesome, start expanding and soon *true friends will come along* and they won't be halves either. How foolish some mortals can be.

Once in a while, two people are brought together by the Hand of The Almighty to do a work together for *Him* and *His honor* and *His glory.* As you learned long ago, how can two walk together unless they agree? You have heard the terms "one mind", "one purpose", "one heart". There is a rare time when two individuals make the conscious choice *together* and *before the throne of God* to join their hearts for all eternity. This is a sacred vow which is made by both parties *in the presence* of the Spirit of God and by His Hand. Such was the choice of your great teacher and author of this book, Dr. Stranges and the editor, Julie. *No thing, no one can ever separate them.* Believe me when I tell you that the forces of Satan himself have made every attempt to do so. Time and again foolish people have attempted to come between them. Even as you read these words, the dreadful, sometimes devastating, actions of certain jealous individuals continue in an attempt to thwart the union between our beloved Dr. Frank and Julie. God Himself has truly blessed their bond. This is a rare and lovely thing. Do not fool yourself into thinking that this is for everyone. Many would rather believe a lie and be damned than to be brought to the *whole truth.*

Asserting Your Greater Self

It is foolish to "give up" anything that will not benefit you in the future. Many "give up" enjoyable things and become miserable. This is absurd. Why not exchange your negative for a positive? Don't give up ill health. Build up good health. No negative emotion can be carried into the coming age of peace. Cease all wasteful financial habits. Discontinue all habits harming your physical body. Stop dissipation in the social department. *Expand the spiritual* through five-point development. Watch your thoughts as well as your actions. Exercise zeal, fervor and enthusiasm in all you say, think and do.

Beware of those who "see good" in everybody instead of seeing them the way they really are. *Advanced* individuals see these qualities but do not dwell on them. Most individuals who recently took their first steps on the Path have been engaged in considerable study regarding the greater and wonderful life to come. Much which is going around today is little better than psychological junk, metaphysical trash. There is an element of truth in just about anything you can mention, yet if you would look forward, how much better off are you going to be when you have finished? Take great care regarding groups, organizations and individuals with whom you associate on a regular basis. Some people fear losing their friends, being alone and spoiling their "fun" if they make the choice to leave a group or disassociate themselves from an organization or individual in order to begin or continue advancing. This is a mass-minded idea. Write out this idea using the exercise which has been given to you within the pages of this book. Your guardian angel has not the slightest doubt about what The Creator intends you to be. For right before you is your *prototype.* He/she protects you from evil and directs you into becoming exactly what God intends. (With your own help, of course) Through your guardian angel you receive tremendous light and revelations regarding the person you are destined to be. Be wary of spending your time trying to discern the hearts and minds and thoughts of others. It is a waste of your valuable resources. Through your own free will you are directly connected with all of the *power of the universe.* Does your heart beat to the same universal rhythm as that of the Universe?

Strange Forces

Many have received the "call" to advanced teachings as did the characters of Mr. Miles, Michael and Mr. Whitehead and continued Spiritual development, yet only a few have accepted the challenges which this call presents. A definite decision must be made and "arrangements" have to be considered. All this usually takes place in a very short period of time. Remember, opportunity presents itself sometimes *only once.* How many chances have you seized and how many have slipped through the holes in your thinking?

You are under no obligation to the teacher but owe much to your fellow human beings. Know the universal law which says, *the greater the individual the greater the obligation!*

When you are faced with an opportunity (problem), why not turn over the dilemma to that Spirit which knows all things from the end to the beginning instead of worrying and wrestling? Relax and let the true Force of Life flow within you and you will get an answer. When the White Forces open a door, only the individual can prevent him/her self from passing through it. The dark forces will do everything in their power to prevent one from taking that step across the threshold of this opening. Do not be caught on the wrong side of the door.

You may be amazed to find how many answers there are just waiting to be realized when you finally come to a firm decision to act. Perhaps you will find that all your worry is for nothing. Be ready to perform in whatever manner is required no matter how "menial" a task it may seem to you. Remember that you are all part of one body and when even the smallest is "out of sync", the entire body suffers. So it stands to reason that even the slightest act of assistance is truly a very large contribution to the whole.

Keep your conversation on a high, constructive plane. Have you ever considered the value of speech? Many have spent hours of time devoted to silence and this training has brought the value of speech forcefully to their attention. After silence, one should use words with care and discrimination. The old adage, "sticks and stones may break my bones but words can never hurt me" is far from the truth. Words can hurt and words can lift up the hearts of individuals. *Never say* to someone what they *do not need* to hear at a particular time, especially if it is of a negative nature. Choose your words carefully.

Expansion

It is the responsibility of some select individuals on Earth to be appointed "guardians" of an area of land or a group of people. It is their duty to "supervise" the great work of five-fold expansion of the dwellers of the area or the members who comprise the group of people. All these individuals are well-trained in the mysteries of the Universe and are Spiritual men and women who have been specifically chosen for their assignment. Many methods are employed in order to accomplish the task for which they have been commissioned. The written word, audio and visual means and gatherings, where the "mysteries" they have learned can be shared with others who by their own choice have selected to become a part of something bigger than themselves, are a few of the ways that these guardians look after their people. The advent of new methods of technology has allowed the ways and means of sharing information to expand very quickly. And many of these avenues are being utilized today by these custodians of information. You would do well to avail yourself of as many of these processes as possible if you wish to advance quickly on your own Path.

It is *not* the duty of these "guardians" to provide every answer to every single question or problem which arises in your everyday life. These individuals are guardians of knowledge and it is their duty to bring forth this knowledge to those who will listen and learn. For it is this quality of listening and learning which will determine how far and how quickly you will be taught and go forward and upward. Are you truly learning that with which you have been entrusted in this book as well as other newsletters, journals and writings which you have been instructed to read? What are *you* doing with the knowledge which has been given to you here and by other means? Are you making it a part of your daily *living* or it is merely stacked on the shelves of your mind accumulating dust and cobwebs? Please read and study in *Outwitting Tomorrow* the chapters regarding the mind. Remember that knowledge and facts are not good for anything but clogging the mind unless they are put to use in your daily life.

Aboard Victor One, Vice Commander Teel occasionally conducts a class entitled *Physio-Psycho Therapy* to individuals who are visiting from planet Earth.

You already know that the human body, in fact every living organism, emits an "aura." These auric emanations from a person seem to come from the entire body, particularly the head. Often they appear in the most beautiful colors. The radiating auras among those who have advanced far on the Path are gold, yellow, brilliant blue and bright light-red with traces of green. Many human auras have a considerable amount of brown, black and gray. Brown indicates things undone that should be done. Black is the absence of all light and indicates underdeveloped possibilities and the extent to which the dark forces have power over one. Gray is indicative of a "doubting Thomas" (remember the story in Scripture?) who would like to believe in higher things but does not have the faith, will-power and desire to do so.

Atheists, agnostics and infidels are walled in with a gray fog-bank. They are cold, chilling and repelling. This gray poison is of their own making and only they can dissipate it. Do you have even wisp of this terrible gray death in your make up? When you begin to doubt yourself, this infernal gray creeps into your aura and begins to expand. If allowed to envelope your pure thinking, this fog bank will surround you and freeze you in your tracks leaving you unable to think properly and move ahead on the Path.

All mass-minded members of the human race as well as evil and ignorant people have much of these three colors in their auras. It is more than the absence of light. It is as a dense vapor or smoke, like soot from a chimney. Not only evil people unknowingly surround themselves with this sooty blackness, but very honest, upright people emanate it when they begin to doubt some Spiritual Truth which they *know* in their hearts to be a fact.

Many have asked, "How do you know there is anything to the idea of auras? Isn't it possible that it is a figment of one's own imagination?" There are machines with which one can view the human aura. Some people are even able to see them with their own human eyes. This ability can be either positive or negative, good or evil. Do you understand why? When one places his "faith" is such things rather than in The Creator who leads and guides us all, this "ability" crosses over into the realm of darkness. On the other hand, those who recognize this capability as a Spiritual "gift" from above, will continue to exercise it for the benefit of human kind.

Humanity is in all phases of expansion. No two persons, whether mass-minded or individuals already on the Path, possess exactly the same expansion in all five departments of life. There are often great differences between people when they are viewed from the perspective of the auric emanations. At this point in time, these visions are opening by leaps and bounds and some are also being used for evil purposes. It is one thing to see the human aura and it is quite another to read it. For example, it is one thing to see the page of a foreign language book but it takes time to learn to read and understand its meaning. It is not important for you to develop such sight. When the time is right, inner sight will be opened naturally. Do not become carried away with fascination regarding this phenomenon and neglect your true path and advancement. Remember also that the *Ring of Fire* is for your benefit! Now it is time for you to again recall the *Ring of Fire* ceremony prayer. The first prayer was given to you for your protection from outside forces. This is the prayer for the express purpose of expanding your inner (Spiritual) self, and it bears repeating here.

'Eternal Father, creator of the universe,
hear this hour my earnest prayer.
Anoint my ears, my lips, my mind and heart,
with your divine flame of fire.
Provide me now with
perfect health, abundance and longevity.
Revitalize and redirect my energies into divine service.
Touch me with your divine inner flame of fire and
cause me to be the master
that you intend me to be.
In the blessed name of the master and creator of the universe,
The Lord Jesus Christ,
so be it,
Amen!'

Recite this prayer at the start of each day to open your heart for what is intended for you to make yourself ready to fulfill that particular working day. Envision in your heart and in your mind's eye (using the Mind of God) that which you are to accomplish today, then follow the dictates of the Seat of God within your body (your heart) and go out and continue to perform the works which God has called you to do.

The Mind And Your Health

The mind is a very powerful "instrument" and can be used for good as well as evil. The negative use of the mind by sinking into self-pity, fear, worry, anxiety, jealousy, hatred, anger, revenge, even timidity and bashfulness can result in illness to the body. Read again the chapter in *Outwitting Tomorrow* regarding the psychological relationship to many illnesses. Even the scientific/medical community is having to admit that there is a relationship between the mind and the body.

The expanding individual seeks knowledge and wisdom in *all five Departments of Life.* There are some medical practitioners who are expanded and possess the necessary knowledge to truly assist you in improving your overall health. They never go about forcing their good works on others and *never* announce that they can heal someone. They confine their treatments to those who are really interested. Anything *you* can do for the mass-minded or negative person is at best only temporary because they haven't and probably will not remove the negative emotions or thoughts which could be causing the trouble. Be tolerant of their ignorance if they choose to keep it. Remember, there are physiological causes for many diseases and disorders. Do not begin to presume that you are capable of "practicing medicine" because of your increase in knowledge. You just may find yourself in more trouble than it's worth. Be wise in your ideas and words.

A Reminder

You must continue to develop your reason, will and judgment. Many responsibilities have been given to you because of the knowledge and teachings you have received. How have you responded to them? Are you making them part of your daily living? Have you taken advantage of opportunities which presented themselves to you? Where are you on your own Path now as compared to yesterday? Are you making forward progress, stagnating or falling back? Have you made the *ring of fire* prayers a part of your daily routine? Only you are able to answer these questions for yourself. Speak honestly to your heart and you should you find any of your own shortcomings. Decide *now* to put into action those changes which will result in your walking tall and quickly on your Path of *expansion.*

Rid Yourself of Evil Spirits Once and For All

Many of you are familiar with the term "exorcism" either from the movie or perhaps through your formal religious training.

An exorcism is the act of expelling an evil spirit, usually by a religious or solemn ceremony. I have been given a prayer that each of you can recite to displace these horrid entities, to exorcise evil spirits.

Should you find yourself in the position of needing to rid yourself of some undue influence, recite the following Prayer of Exorcism of Evil Spirits:

In the Name of the God of Everlasting Creation
I, as a child of Almighty God hereby command
all evil forces of darkness to release me
from every negative force, influence and authority
that is binding me and preventing me
from being fruitful and successful.
I hereby rebuke all of the forces of the evil one.
I command you to take your hands off me, now and forever.
I now exercise full authority to bind and cast away from me
all evil influences.
In the Name of Almighty God,
Our Lord and Master,
Jesus Christ.
Amen.
So be it!

Appendix B

Why Physical-Financial Success Doesn't Pay Off

The physical department of life is ideal for getting a financial start, but from there on, it is out as far as financial success is concerned. Most successful people work with the physical body to secure enough money to get started in the financial world. But then one must use the mind. Save something, no matter what your wage. Look for investments and use good judgment in placing your money there. Never go into business without a thorough investigation. The subconscious knows how little you know about an undertaking and it may attempt to separate you from the new business venture thinking it is doing you a favor. So this can't happen, take a tremendous interest in the new business and learn every angle just as quickly as possible. If you're interested and desirous of making a fortune out of this business, your subconscious will then understand that you *do* want to receive a great benefit from it. Do this and success is yours.

Financial Increase Caused By Spiritual Expansion

When it comes to your star of development, the Spiritual department is directly connected to the social. The financial is connected to the mental department wherein lie *reason, will* and *judgment*. The mental department also contains the subconscious, emotion, affection, belief and genius. Intuition, inspiration and genius are at home in both the subconscious and the superconscious. Keep in mind that the *superconscious is directly related to the Spiritual department.* The more Spiritual one becomes, the wider the channel between the superconscious and the conscious. Employ zeal, fervor and enthusiasm is every one of your activities, especially your giving and the first thing you know, you will be receiving. Be grateful for everything that comes to you. It is there for a reason.

Wealth is the Shortcut to the Evergreen Life

The mass-minded human beings are like trees that grow old, turn yellow and die. Not so with an individual traveling upward toward evergreen life. Money vibrates to green. Sap, the blood of plants, flows upward giving a constant, delightful state of youth. But the moment it stops, the tree stops *expanding*.

What good would it do if you made the world a better place to live? It would be entirely peopled by zombies. The human race has a great destiny planned for it, wonderful beyond the imagination. In fact it has already begun, especially for those who have made the decision to turn "west" and follow the Path upward and outward. For this reason, *destiny will not permit* any idealist to stop the progress of humanity by making life so wonderful that the human race will stagnate and die thinking there is nothing further to achieve. If you're an idealist, trying to make the world a better place to live, stop your stupid activity at once, turn about face and start making the human race, member by member, one at a time, better. It can't be done en mass. Do your share of guiding people into the five-fold balanced path of life. *First make yourself the ideal person* Destiny intends you to be.

It doesn't matter how old you are, once you begin to *expand* your financial department, it will have a magical effect on your mental and Spiritual departments. The moment you cause the green financial sap to flow upward in your financial tree, the physical and mental trees of life will immediately show signs of springtime. There is but *one* tree and five branches representing the five departments of *life*, also known as the *Tree of Life*. (It is worth noting here that this is the same tree of life that existed in the Garden of Eden from which Adam and Eve were banished because of their disobedience in eating from the tree of the knowledge of good and evil.) If a person devotes one's entire time to things financial, one becomes a lover of money. This causes the other four branches to wither and die. Could you see yourself as a tree? The average individual would see a mighty poor tree with the financial branch and the physical branch faded and yellow. The social branch is seen in grotesque shapes and the mental branch as unstable, nervous and ever-quaking. The Spiritual branch would be the most amazing. It would show how the financial sap once made the branch beautiful during the up cycle. Then the down cycle would show the old dead branches with a slight show of life.

Your financial, Spiritual and mental departments must work very strongly together if you are to *increase all five departments*. Force yourself to take an interest in the accumulation of wealth. *Like the idea* of accumulating sufficient means to accomplish the goals you have set for yourself. Be assured that this is not being selfish. You can't fool the subconscious. It will catch on to what you really want.

Most people want to receive first, then give. It doesn't work that way. Make it the other way around and you break the most damnable financial cycles, regardless of your age, environment and circumstances.

Why Every Master Possesses Great Wealth

To become a master, one must possess the *ability* to have unlimited wealth. This is necessary for the great work that is to be accomplished in the future. There's a difference between possessing the *ability* to accumulate great wealth and possessing great wealth. Being older gives one the freedom to roam the Earth doing whatever they are called upon to do to help others along the Path. When a true master is assigned work that requires a great deal of money, in some cases one doesn't have a bank account. One simply makes him or herself valuable to some business concern, corporation or establishment to the extent of the required money. Then one picks up their check and goes about their work. Whatever one needs, a master simply dispenses services worth that amount.

The expanded mental department of the masters brings to them the solutions of others' relatively simple problems. Any number of masters are engaged in business activities. All the money they can use, they have. There is no need to materialize money as do so many of the clever black masters as a demonstration of their powers. Masters raise money in a dignified, business-like way and have no retribution to catch up with them. Five-point expansion really "pays off."

The Trinity of Human Accomplishments

A most important and effective tool for your Spiritual and human advancement is the recognition of the value of reason, will and judgment. A complete understanding of these can be reflected not only in the contributions of all the forces of good, including the angels, but also in Holy Writ. Without sermonizing, please allow me to expand on these three. Study and absorb the pages of this chapter very carefully and pray that your pure understanding will be enlightened for your own personal benefit.

Reason

Reason is defined as, "the mental powers concerned with forming conclusions, judgments, or interferences; normal or sound powers of the mind; good sense" (according to Webster's College Dictionary, 1995) and can be rightfully understood in the context of direct application. For example, there are actually two forms of reasoning. The first addresses itself directly to the human brain and the second applies directly to the Spiritual. In many cases, the two work together in perfect harmony and concert. Let us first examine the human aspect.

Human reasoning deals directly with the various, countless neuro transmitters, lights and beams of electricity that formulate our brain. There are more connections working in the small portion of our brains which we use than in all the computers combined. When our brain is working to capacity (as much as it is today even though it's a very small percentage of what our brains are capable of) our powers of human reasoning operate without becoming short-circuited. Our thoughts and ideas appear fresh, simple and uninhibited. Sound human reasoning results when one is in full control of all of his faculties.

A man or woman who is of sound mind is capable of making snap decisions at the drop of a hat. Sound thinking and solving a problem pose no obstacles whatsoever. Human reason demands that the individual knows the road upon which he or she is traveling. They have already formulated this premise in the deep recesses of their brains. Remember that the mind is not only contained within the confines of the physical brain itself, but also partially exists outside the physical and which is the Spiritual portion of the mind. Therefore, reason can be realized and appreciated by others. The goals, aims and purposes of each individual have already been defined and established.

There appears to be an unquestioned, positive, rational human reasoning when one of sound mind arrives at a decision or a conclusion. Human reasoning comes into play, the decision is made and acted upon without hesitation, delay or apology. The powers of human reasoning are called to the forefront and the individual finds that their decisions are not only quick but also accurate. When human reasoning is summoned, individuals do *not* act rashly or carelessly or actively engage in anything which would betray their powers in any way, shape, manner or form.

Spiritual reasoning, on the other hand, represents a higher form of enlightenment. This comes as a result of the individual coming into contact and communication with the Divine. Spiritual reasoning comes as a result of obeying the voice of heavenly intelligence whose outcomes are amply reflected in one's works, acts and deeds. When dealing with the Spiritual aspect on one's life, the individual finds him or herself in a far different scope of influence than in dealing with the human. The Spiritual application to one's life produces positive results in place of negative.

When one finds that the human mind is literally inspired by those higher Spiritual powers, one becomes aware of walking on a much higher plane of existence than before. You can apply these Spiritual rules of conduct to your pure powers of Divine reasoning and find that the results are more than you as a human being ever expected. Working in concert with the Spiritual mind produces powers that will cause you to walk and talk like the being that The Almighty intended you to be since the beginning of time. Once you have acquired the Spiritual connection to the "Mind of God", or the "Mind of the Spirit", your life begins to change from a "clod to a god." Your actions, decisions and accomplishments amply reflect the irrefutable truth that you have spent time in the presence of the Divine.

Judgment

Judgment is defined as "the ability to make a decision, or form an opinion objectively or wisely, especially in matters affecting action; good sense; discernment." (Webster's Collegiate Dictionary, 1995) It is almost impossible to separate *reason* and *judgment.* They directly affect each other. In many cases, they are dependent upon each other. Judgment is that singular power that is the underlying reason for success or failure. How many times have you heard that either this person or that made poor decisions in life resulting in utter disaster? This can be applied to business, associations with others, purchases, sales and also marriages. It is the supreme desire of the Almighty that you, as an individual, seek to elevate your pure thinking to the point that the judgments you make in every area prove themselves to be fruitful.

Judgment is a most important single factor in your overall human makeup. Poor judgments have proven to be primarily responsible for negative decisions made in life. Poor judgments have also caused many to experience an untimely demise. Judgments (decisions) regarding your diet, your choice of friends, business decisions and associates, your mode of travel plus your overall human practices and the like must be carefully considered. Good judgments often pay great dividends. Many times, good judgments are noticed by the world in general.

You are automatically placed in a special category when you engage in sound judgments. There are those in the world (whom I call "the watchers") who do in fact watch, observe and judge every move that you make with peering eyes and a desire to know "what's going on in your personal life." Once you are elevated to a higher pattern of living because of your sound judgments and wise decisions, you will begin to think that you are living in a glass house. Along come those watchers who pace your every move. They observe, they watch and they judge you for better or for worse. Do not allow yourself to fall into the trap of being a watcher who sits in judgment of others and "decides" what is happening in the lives of others while at the same time neglecting to sit in the seat of judgment on yourself.

Will

Will is defined as, "the faculty of conscious and particularly of deliberate action; the power of choosing one's own actions; the wish or purpose as carried out or to be carried out." (Webster's Collegiate Dictionary, 1995) The will can likewise be recognized as having a twofold aspect. One is the Divine and the other, human. First, we will examine the human.

The human will is responsible for many blunders throughout history. When one wills to do a thing without thinking, it often results in a catastrophe. It has been said that historical figures have acted out of compulsion because of a strong will. History reveals those who exercise a strong will do not believe in consulting others. And even if they do, it is only for show because they have no intention of following the advice of others. Their will is so strong and they are very set in their ways of decision-making. It was not important to them whether their particular strong will was right or wrong. Leaders of nations dating back thousands of years will be remembered because of the manner in which they used their will. The human will creates many difficulties and unnecessary problems for those who are not ready, willing or able to listen to reason. Many historical leaders have later realized that they had unwittingly plunged themselves into situations without first thinking things through from all aspects of the situation.

I believe that by now, you notice that there is rather an interesting overlapping of reason, will and judgment. All three address themselves to the intricate affairs of the mind as well as the heart. The will of man is often set in stone. When a proud man sets his course (his will), very little if anything can cause him to change his thinking on any given project. He will embark upon adventures that have tickled the fantasy of his own selfish motives (will). There are thousands of men and women, the world over, who have freely displayed the full exercise of the human will with varied results. And unfortunately for many, some of these have been world leaders.

Then again, we must not exclude the Divine Will of Almighty God working in concert with the human will of man. When the time should arrive that a person's understanding becomes enlightened, one will find that he has accomplished more in life because he listened to the Spirit of Truth while exercising his own human will at the same time. He was able to come up with the perfect blend of Divine and human wills. In order to accomplish this, one must learn the valuable secret of self-discipline. There are times when the human part of us wants to run away. On the other hand, the Spiritual will wishes us to retreat and reexamine our thoughts and intents. As it states in the Biblical records in the writings of Solomon, there is a time to act and there is a time to stand still.

Solomon further states that there is a time for everything under the sun. When we exercise the Will of God within our own minds, we find that there is a lighted Path for us to follow and we are not stumbling around in the darkness of ignorance, fear and superstition, the darkness of indecision and despair. The will of mankind can be used for good or for evil. The will of man, working with the Spirit of Truth, will prove to be a union of which there is no parallel in the universe.

This important lesson is already being practiced by those from other planets (including Angels) who did not lose their first estate. In other words, they remained faithful to the One who created them and did not choose to become part of those who were cast out of the heavens and into the Earth, and who are now known as the dark forces following a great war among the angels. These are the same dark forces referred to in the main body of *Outwitting Tomorrow.*

When the human will is completely surrendered to the Divine Will of Almighty God, one enters into a state of being whose power and influence can change the course of human existence on this planet forever. Therefore, learn and appreciate that these three, reason, will and judgment, represent a wholesome trinity that if wisely combined can and will permeate a person with understanding, wisdom and knowledge which will most certainly set you aside as one of His elect. You will then be in a position to set your feet westward on the Path and travel with lightning speed toward those specific aims and goals which Destiny has set for you and that will insure you the prosperity, good health and longevity which you deserve and for which you have long awaited. Combine zeal, fervor and enthusiasm with reason, will and judgment and using the Mind of God in conjunction with your own human mind, you will be on your way to a life filled with all the joys that traveling the positive Path can bring to you. You will be astonished at what you can accomplish in your own life if you apply all the lessons which have been given to you. Study the significance of the drawing referred to previously as the 'Circle, Triangle, Star'. This simple drawing illustrates the main points which you should remember. And if you do, truly, you will be ready and capable of *Outwitting Tomorrow.*

About This Book and the Author

The original manuscripts of *Outwitting Tomorrow* were given to the late H.J. Gardner in 1932. Part of it was published under the pen name of "Frater VIII" in 1939. Evidently the story was ahead of the times, its profound message uncomprehended. The original printing house is no more, the plates have been lost or destroyed.

This new book is written by Dr. Frank E. Stranges, with Commander Valiant Thor serving as an advisor and is being presented for the first time here.

Dr. Frank E. Stranges and Valiant Thor are becoming well-known to ever-increasing segments of the world public. (Read *Stranger at the Pentagon* and *Millennium VII*, by Dr. Frank Stranges, available from the publisher or author of this book.)

The progress of the human race on Earth is of intergalactic concern. A plan has already begun to awaken the human family to *reality* beyond the end of its nose. In the recent past, four seven-year periods were allotted for attempting arousal of *awareness* in major areas of human endeavor. They were the scientific, political, academic and religious centers, in that sequence. Dr. Frank E Stranges, Valiant Thor and several of their colleagues systematically contacted world leaders in each separate category, only to have their offers of information universally rejected. While some actually expressed their belief in the knowledge, the gift of love and knowledge to help break the grip of the dark forces which are strangling this planet and its people was refused by each of the sectors contacted. Various reasons were given for the refusals, mass hysteria and complete disruption of the economy (guess who); loss of their congregants; harassment by their colleagues; possible expulsion from their denominations if they were to accept the invitation to relate such teachings to their respective members; outright prejudice toward the information itself.

Time is running out for us, the human family, on Earth. At this late juncture, an organized undertaking is being made to contact and awaken those open-minded individuals everywhere who will consider these simple truths and their application as described carefully within the pages of this book. Dr. Frank E. Stranges is one of those people who was contacted over thirty years ago, was chosen and accepted the challenge because of his dedication to presenting 'the truth' without fear of repercussions from any quarter. He agreed to learn and then teach the Universal Truths which will free people from certain constraints which have held them back for centuries. This he does through various audio, video and printed materials. Humanity is at a curious junction as the new millennium begins and moves forward into eternity. Each person will have to make a choice as to which path they will follow into their own future—the one set forth by Destiny or the one of their own making.

Your own personal destiny as well as that of the entire human race depends upon the choices you make of your own free will and how quickly you make them. These messages will help dissolve the differences among the open-minded and the mass-minded which have deteriorated the minds, hearts and spirits of the human family through lack of action for so long. Ponder the beauty in a world of *expanded individuals*, mightily prepared by these few straight-forward and uncomplicated lessons contained here in *Outwitting Tomorrow.*

Outwitting Tomorrow, by Dr. Frank E. Stranges, has been carefully written in simple terms with advice of a Starship Commander, Valiant Thor, an Immortal among us today. *Outwitting Tomorrow* contains definitive instruction regarding specific subjects related to Universal Laws resulting in immediate positive results for the protection, health, prosperity and well-being of those willing to apply them. Dr. Stranges or Commander Valiant Thor may be contacted directly through the publisher or via email at drfes@earthlink.net

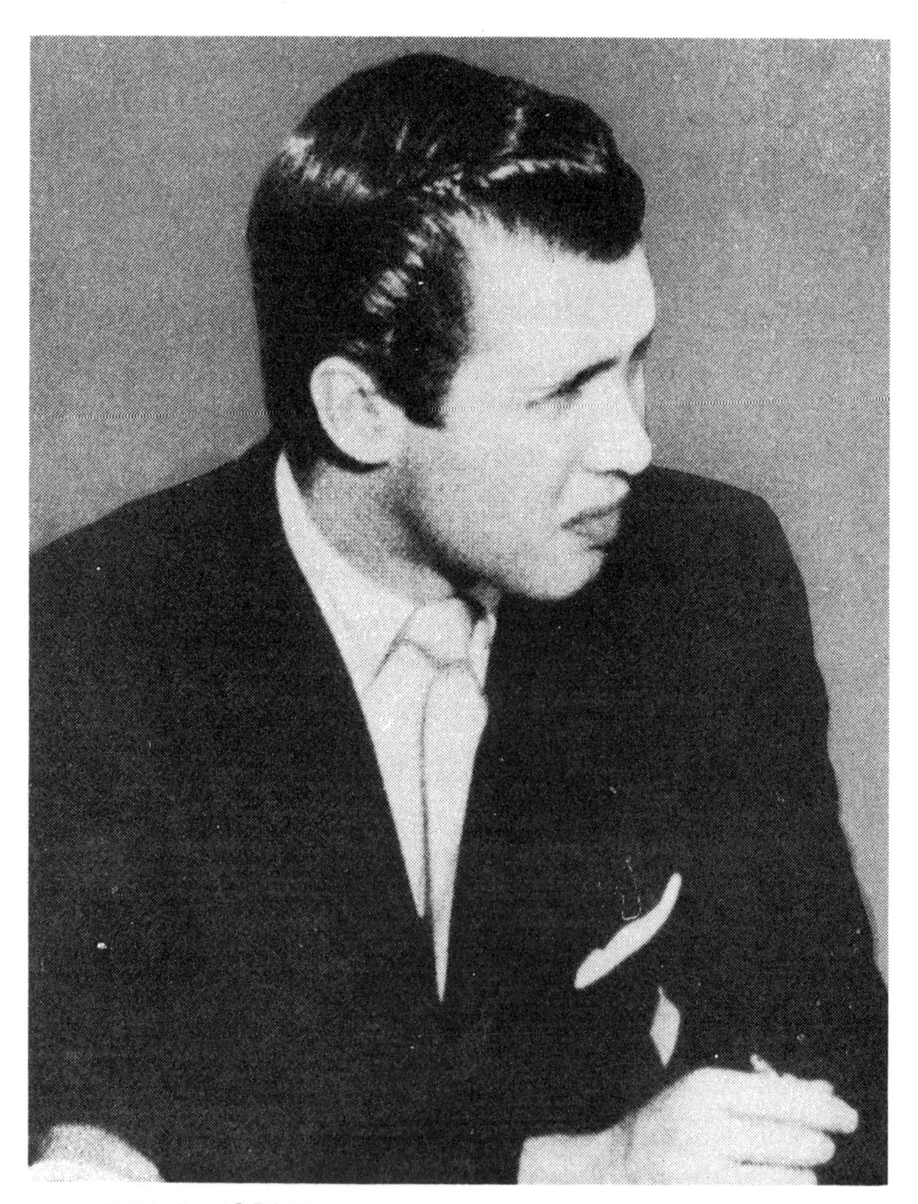

Photo by August C. Roberts

Valiant Thor, Advisor to Author

Any similarity of the above photograph to persons living or dead, is purely coincidental.

Footnotes

[1] Height of the Great Pyramid—481 feet. Each of the four baselines, 760 feet, 11 inches. Distance around the Great Pyramid, 3043 feet, 8 inches. Diagonal base measurement from the corner to opposite corner—1076 feet, 1 inch. Base of the Great Pyramid covers approximately 13 1/4 acres. Each side, 5 1/2 acres. All four sides, 22 acres. Approximately 90,000,000 cubic feet of stone used in its construction. Composed of nearly 2,300,000 individual blocks of stone. Average weight of stones, 2 to 2 1/2 tons each. Largest stones weigh approximately 30 tons each.

[2] Worked into the construction of the Great Pyramid of Gizeh are the following (1) The weight of the Earth. (2) The exact sphericity of the Earth. (3) The mean density of the Earth. (4) The exact length of the Earth's polar axis. (5) The direction of true north. (6) The Earth's orbital maximum variation. (7) The mean distance from Earth to sun. (8) The Earth's mean temperature as indicated by the air temperature in the King's Chamber. (9) The exact British and American inch, foot, yard, furlong, and mile. (10) The exact British and American grain, ounce, pound and ton. (11) The standard British and American pint, quart, gallon, and bushel. (12) The practice of squaring the circle. (13) The practice of quadrature of the circle. (14) The exact center of land area of the Earth. (15) The cubic capacity of the open Treasure Chest in the King's Chamber is identical to the Ark of The Covenant. (16) The capacity of Noah's Ark was exactly 100,000 times that of the open treasure chest.

[3] Events foretold by the Great Pyramid—Beginning of the "Second Creation" (Adam), 4,000 B.C. Beginning of the Great Flood (Noah), 2,845 B.C. The Exodus, 1,486 B.C. The Birth of Christ, October 6th, 4 B.C. (Julian Calendar). Baptism of Christ by John the Baptist, October 3rd, 27 A.D. The crucifixion of Christ occurred on Friday, April 7th, 30 A.D. From Pentecost to 400 A.D., the Great Pyramid shows the Christian Church to be spiritual; the church then reached its all-time low spiritual state from 1000 to 1300 A.D. The beginning of the material and mechanical age, 1844 A.D. The beginning of the air-minded age, August 2, 1909. Evil forces in the Bottomless Pit become liberated and active, March 12, 1913. Great Britain's entrance into World War I, August 5, 1914. The first restrictions on the Jewish people, January 18, 1918. Ending of the World War, November 11, 1918. Beginning of the international depression, May 29, 1928. Beginning of the NEW DISPENSATION, September 16, 1936. Beginning of the "one hour" or 15-year period, August 20, 1938. The Bottomless Pit closed to the evil forces and forcing them into the "Dead End"(8X) passageway, November 27, 1939. Halfway mark across the Judgment Hall (King's Chamber), March 4, 1945. Halfway mark of the "one hour", 15-year period- February 18, 1946. Completion of the "one hour", August 20, 1953.

[4] Substitute "the United Nations Forces" for "all nations" in the above verse and see how up-to-date it becomes.